STUDENT WORKBOOK

to accompany

Elementary Algebra

EQUATIONS & GRAPHS

Katherine Yoshiwara
Bruce Yoshiwara
Los Angeles Pierce College

Irving Drooyan

Australia • Canada • Denmark • Japan • Mexico • New Zealand • Philippines • Puerto Rico
Singapore • Spain • United Kingdom • United States

Project Development Editor: Michelle Paolucci
Marketing Manager: Leah Thomson
Marketing Assistant: Debra Johnston
Editorial Assistant: Erin Wickersham
Production Coordinator: Stephanie Andersen
Print Buyer: Micky Lawler
Printing and Binding: Webcom Limited

For more information about this or any other Brooks/Cole products, contact:
BROOKS/COLE
511 Forest Lodge Road
Pacific Grove, CA 93950 USA
www.brookscole.com
1-800-423-0563 (Thomson Learning Academic Resource Center)

Printed in Canada

10 9 8 7 6 5 4

ISBN: 0-534-36540-X

Note to Students

This workbook contains tables and grids for the activities and problems in your textbook. The grids are already labeled with appropriate scales for the graphs you will draw. We hope that providing these grids will eliminate some of the time-consuming work involved in drawing a graph, and allow you to concentrate on the mathematics.

The lessons in your textbook include Exercises printed on a yellow background. There are copies of these Exercises in the workbook, with space for you to show your work and record your answers. You should try these Exercises as you read the text, to see if you understand the material.

You should try to keep your workbook up to date as your course progresses. You will also need a spiral or loose-leaf notebook for class notes and the rest of your homework problems. Your workbook and notebook will be useful study aids when you are preparing for exams.

How to Be Successful in Your Math Class

The key to success in a math class (as in most endeavors) is persistence. You cannot learn mathematics in one great rush the night before the exam; but you can master it in small chunks a little at a time. You should plan to study math for at least one hour every night. Don't give up until you have a good grasp of the lesson and can work the problems on your own. If you get behind in a math class it is very difficult to catch up.

1. **Attend class every day.**
   Studies have shown that success in math classes is correlated strongly with attendance. If you must miss a class, find out beforehand what the class will cover. Read the lesson and complete the assignment anyway, just as if you had attended.
   a. Use class time wisely. This is your best opportunity to learn the material.
   b. Take notes. Learn to summarize what the instructor says, not just what he or she writes on the board.
   c. Don't be afraid to ask questions when you don't follow the lesson.

2. **Read the text book.**
   Reading a math book is not like reading a novel. You will need to read the material more than once to understand and retain it.
   a. Read the new material *before* it will be covered in class.
   b. Read with a pencil in hand so that you can make notes to yourself, underline important points, or put question marks in the margins.
   c. Read the section again after it has been covered in class.

3. **Look over your handouts and class notes.**
   The sooner you can review your notes after class, the better. People forget most of what they hear very quickly, and reviewing your notes will help you retain the new information.
   a. Look for points where your notes reinforce the material in the textbook.
   b. Try to fill in any steps or information you may have missed in class.
   c. Write a sentence or two summarizing the main points of the lesson.

4. **Do the homework problems.**
   Most of your learning takes place when you work problems. If you do some of your work in a study group or tutoring center, you will have someone to consult as soon as you hit a snag.
   a. If you get stuck on a problem, refer to the textbook or your notes for help.
   b. Call a classmate on the phone and try to figure out together the problems you had trouble with.
   c. Mark any problems you can't get, but don't stop! Skip those problems for now, and continue on to the end of the assignment.

5. **Get help right away.**
   Mathematics builds upon earlier material, so if you don't understand today's lesson you will have even more trouble tomorrow or the next day.
   a. Make a list of points you don't understand and problems you need help with.
   b. Ask your instructor or a tutor for help *today* -- don't put it off!
   c. Fill in your notes with the answers to your questions, and make sure you can work all the problems that gave you trouble.

6. **Prepare for exams.**
   In addition to keeping up with daily work, you must prepare specifically for exams. Always study 100% of the material the exam will cover. If you omit some topics, you won't be sure which problems you should work on during the exam!
   a. Begin studying for the exam a week ahead of time, so that you will have a chance to get help on any topics you are unsure about.
   b. Make a check-list or outline of the material the exam will cover, and review each topic until you have mastered it.
   c. Have a classmate or tutor make up a sample exam (or make one yourself), and practice working problems under exam conditions.

Table of Contents

Chapter 1 Variables and Equations

Section 1.1 Tables and Bar Graphs

Airlines	Complaints	Passengers (in Thousands)	Calculations	Results
Alaska	53	10,084		
American	497	79,511		
American Eagle	35	11,900		
Continental	368	35,013		
Delta	504	86,909		
Hawaiian Airlines	31	4776		
Markair	263	990		
Nations Air Express	36	82		
Northwest	257	49,313		
Southwest	107	50,039		
Sun Jet Intl.	142	486		
TWA	291	21,551		
United	597	78,664		
USAir	379	55,674		
ValuJet	83	5145		

Table 1.1

☐ *Use Table 1.1 to help you answer the questions on page 2:*

1. Which airline had the most complaints in 1995?

   Which had the fewest?

2. Which airline carried the most passengers in 1995?

   Which carried the fewest?

3. Locate the data for Nations Air Express and for American Eagle. Which of these two airlines had the worse complaint record in 1995?

   Why did you choose the one you did?

4. Would you say that the airline with the most complaints in 1995 had the worst complaint record?

   How can you use the data to compare the complaint records more fairly? Please describe your method: What computations will you use?

5. Use the last two columns of Table 1.1 to carry out your method. (Round your results to three decimal places.) According to your calculations, which airline had the worst complaint record in 1995?

☐ *Use Table 1.2 to help you answer the questions on page 3:*

6. What is the average life-span for people born in 1930?

   What is the average life-span for people who were 50 years old in 1930?

7. In which decade did the life expectancy at birth increase the most?

8. Is there any decade in which either life expectancy at birth or life expectancy at age 50 decreased?

9. Over the past century, what has happened to the gap between life expectancy at birth and life expectancy at age 50?

   Why do you think this has occurred?

10. Do you think that life expectancy will continue to increase indefinitely?

☐ *Use the graph in Figure 1.1 to help you answer the questions on page 4:*

11. Locate the bars corresponding to 1940 on the graph. Find the heights of the bars by comparing them to the vertical scale on the left side of the graph. Do the heights match the values for life expectancy given in Table 1.2?

12. Look at your answer to question 7. Explain how to answer that question by looking at the graph instead of at the table.

13. In what year did the life expectancy at birth catch up to the life expectancy for 50-year-olds in 1900?

14. By looking at the graph, determine the decade in which the gap between life expectancy at birth and life expectancy at age 50 first narrowed to 10 years or less. Check your answer using Table 1.2.

15. Both life expectancies have been increasing over the last century. Are they increasing more rapidly or more slowly as time goes on?

    Does this change your answer to question 10, or support it? Why?

☐ *Follow the steps in the text to make a bar graph for Table 1.3:*

1.

2. (Use the grid at right.)

3. Answer to question (1):

   Answer to question (2):

4. (Use the grid at right.)

5.

Grid for Bicycle Commuters Bar Graph

☐ *Use the histogram in Figure 1.2 to help you answer the questions on page 6:*

1.

Monthly rent	$0-$200	$200-$400	$400-$600	$600-$800	$800-$1000	$1000-$1200
Number of freshmen						

2. Which of the given ranges of monthly rents is most common?
   How can you tell from the histogram?

3. Which range of rents has more students: $600-$800 or $800-$1000?
   How can you tell from the histogram?

4. Verify that the total number of freshmen polled was 40.

Homework 1.1

3a.

National Origin	US Population in 1980	Increase to 1990	US Population in 1990
Chinese	806	104.1%	
Filipino	775	81.6%	
Japanese	701	20.9%	
Asian Indian	361	125.6%	
Korean	353	126.3%	
Vietnamese	262	134.8%	

3b.

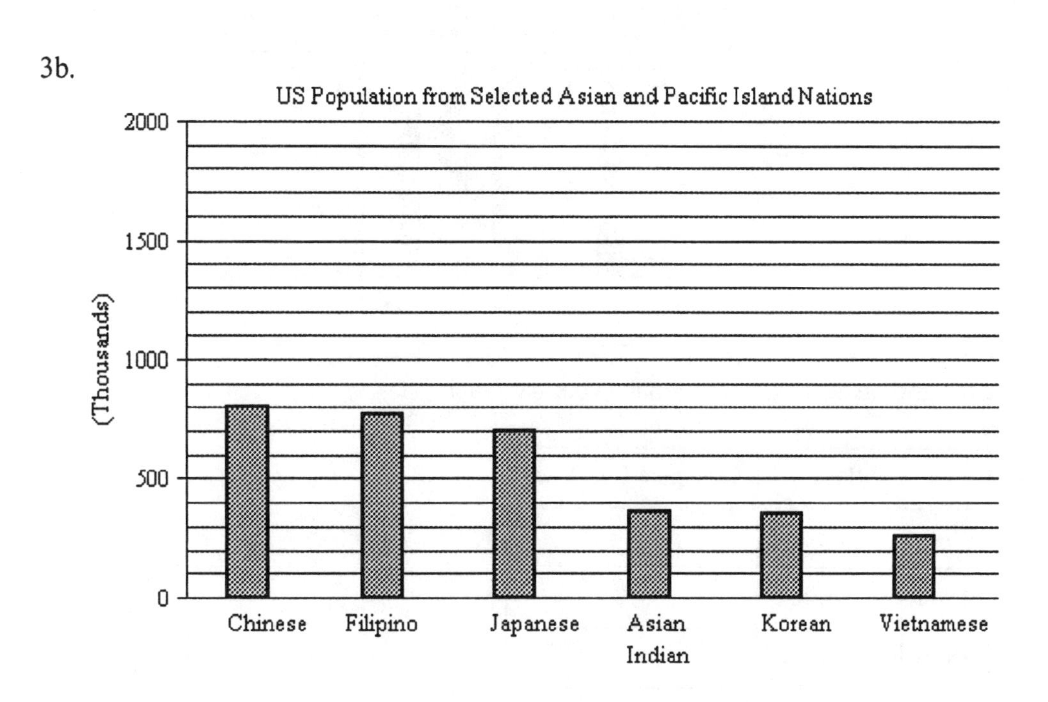

4b.

Sector	1979	1990	1993	2005
Manufacturing				1.1
Trade				1.7
Services				2.6

4c.

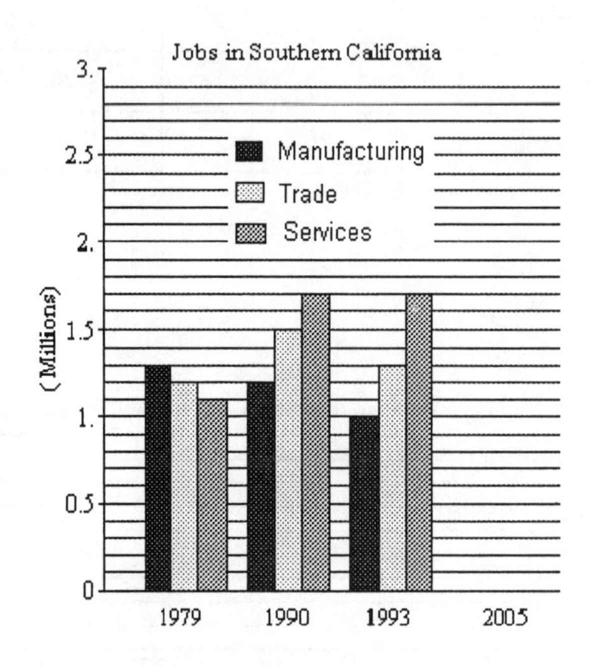

6. (Use the grid on the next page to make your bar graph.)

d.

Year	Salary	Increase Per Year
1940	$8000	-----
1960	$15,000	
1970	$23,200	
1975	$39,600	
1980	$78,700	
1985	$193,000	
1990	$355,000	

Grid for Exercise 6

7a.

Quiz Score	0	1	2	3	4	5	6	7	8	9	10
Number of students											

b.

c.

d.

e.

f.

8a.

Household size	1	2	3	4	5	6	7	8	9	10	11	12	13
Number of households													

b.

c.

d.

e.

f.

Section 1.2 Line Graphs

☐ *Follow the steps in the text to make a line graph of the data in Table 1.10:*

1. Number of tick marks on horizontal axis:

3. Highest price:

   Lowest price:

   Number of tick marks:

7.

8.

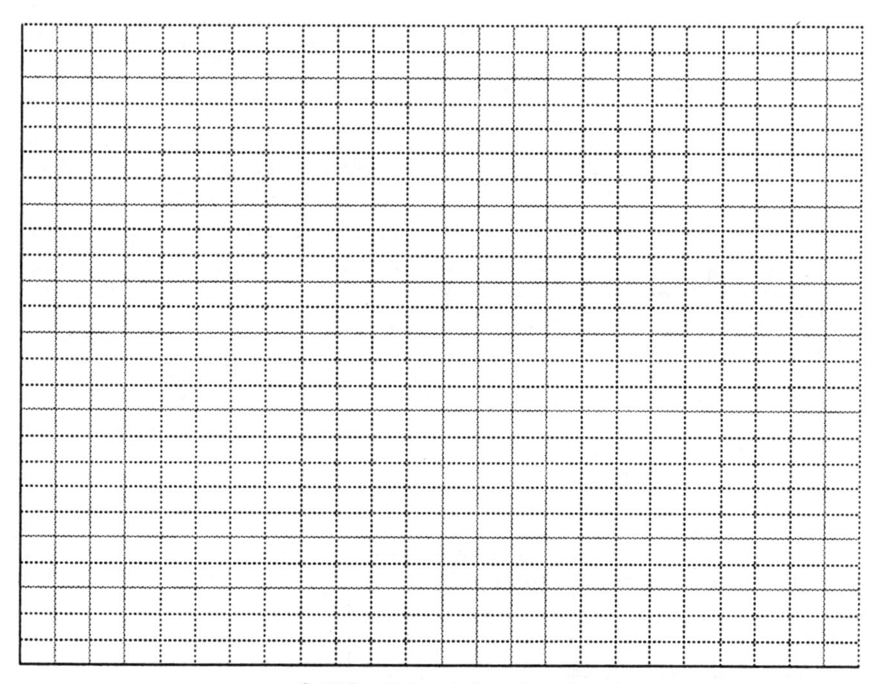

Grid for Natural Gas Line Graph

☐ *Make a line graph for the plane's altitude.*

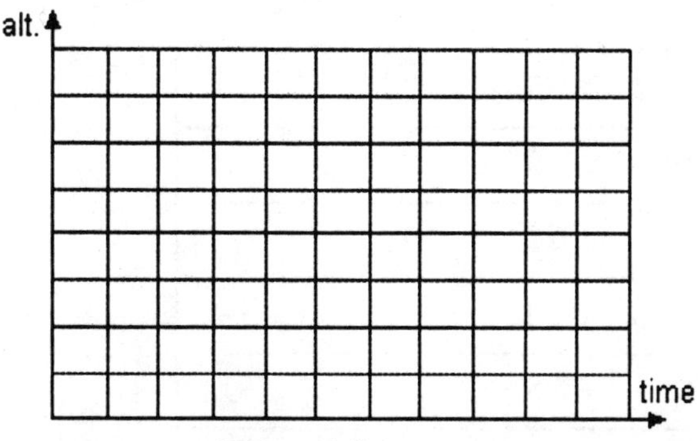

Figure 1.9

Homework 1.2

2.

Age	Height	Growth
0	19	---
1	29.5	
2	33.5	
3	38.5	
4	41.25	
5	43.75	
6	46.5	
7	49	
8	52.5	
9	55	
10	58	

height

age

4.

Decade	Immigrants	Rounded Figure
1901-1910	8795	
1911-1920	5736	
1921-1930	4107	
1931-1940	528	
1941-1950	1035	
1951-1960	2515	
1961-1970	3322	
1971-1980	4493	
1981-1990	7338	

6.

Size of Group	10	20	25	30	40	50	60
Probability	11.7%	41.1%	56.9%	70.6%	89.1%	97.0%	99.4%
Rounded Values							

11.

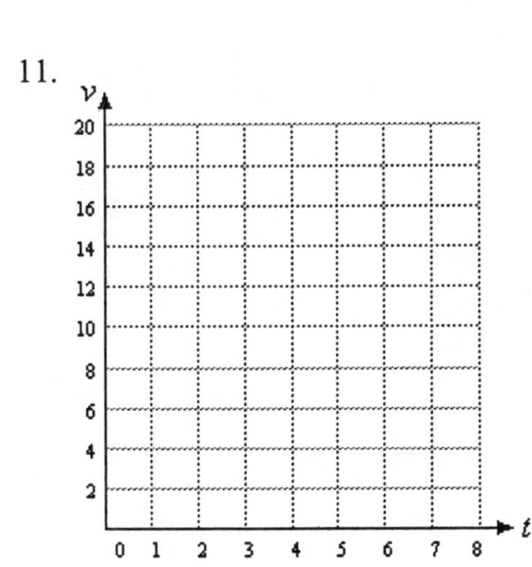

12.

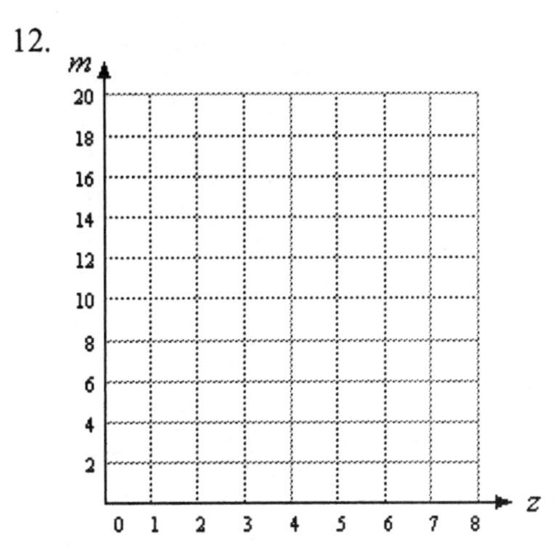

13.

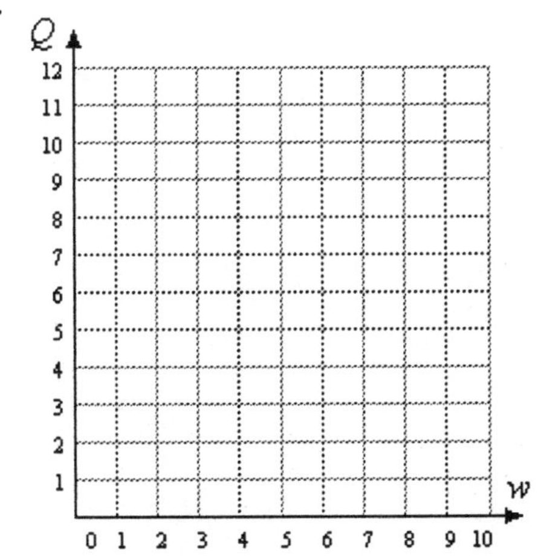

14.

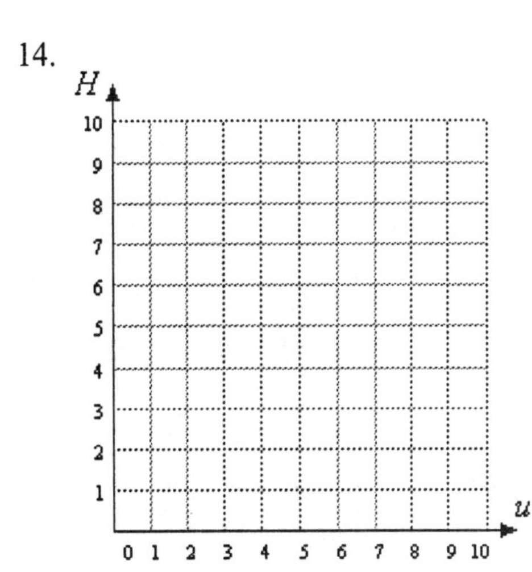

Section 1.3 Variables

1.

Barry's paycheck	45	60	75	100	125	p
Calculation	$45-20$					
Amount he keeps	**25**					

a.

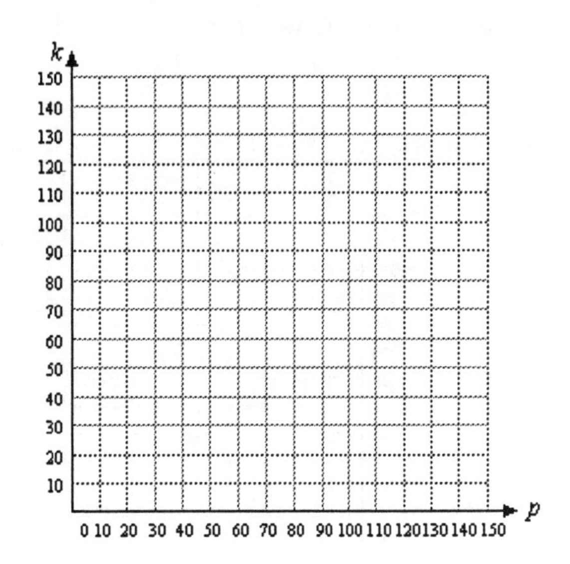

b. Amount he keeps =

c.

2.

Hours worked	3	5	6	8	15	h
Calculation	6×3					
Wages	**18**					

a.

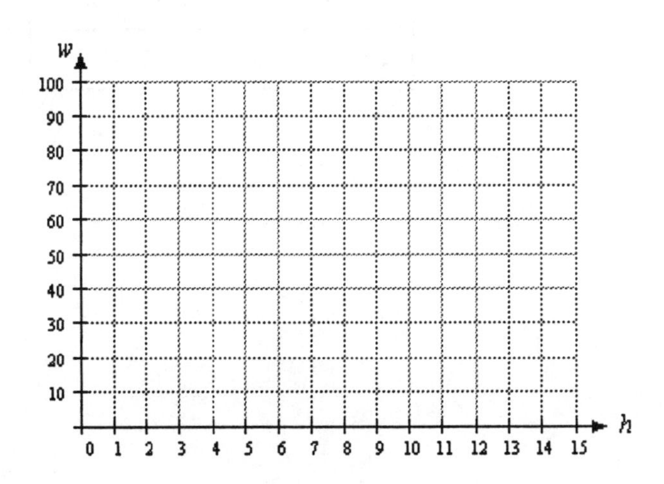

b. Wages =

c.

3.

Ralph's weight	150	165	180	195	210	*R*
Calculation	**320−150**					
Wanda's weight	**170**					

a.

b. Wanda's weight =

c.

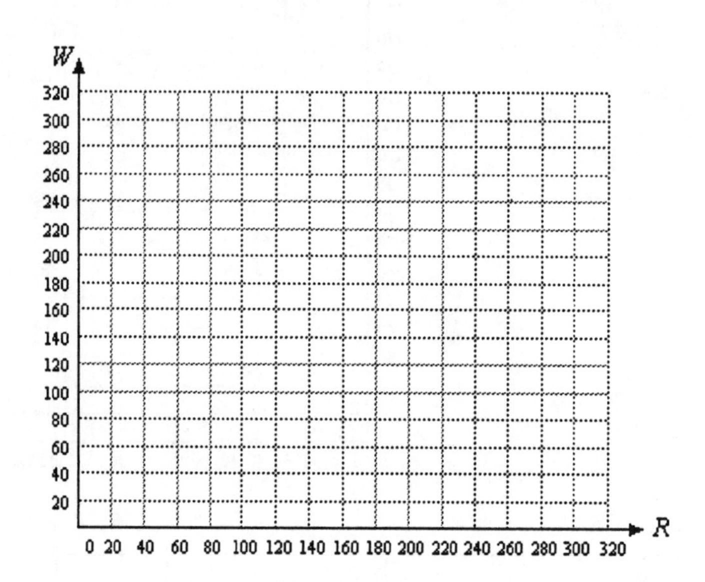

4.

Total calories	1200	1500	2000	2400	2800	*C*
Calculation	**0.30×1200**					
Fat calories	**360**					

a.

b. Fat calories =

c.

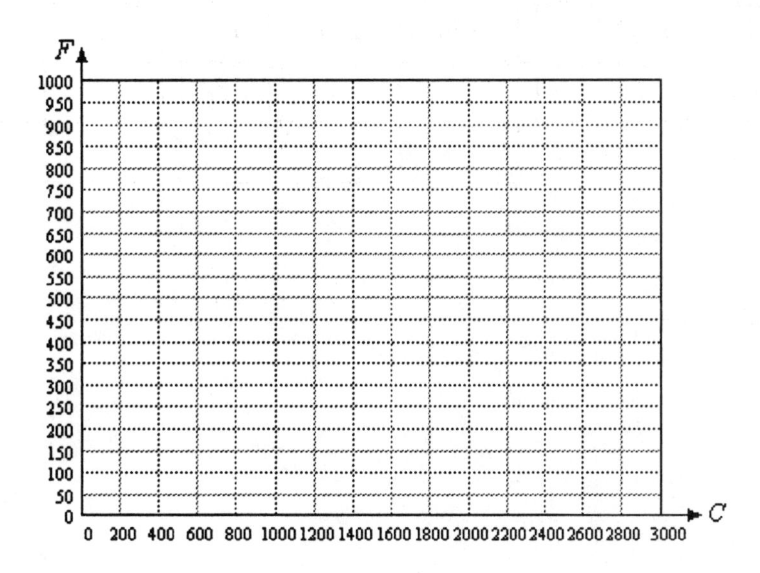

Homework 1.3

1.

Temperature in Ridgecrest	70	75	82	86	90	R
Calculation						
Temperature in Sunnyvale						

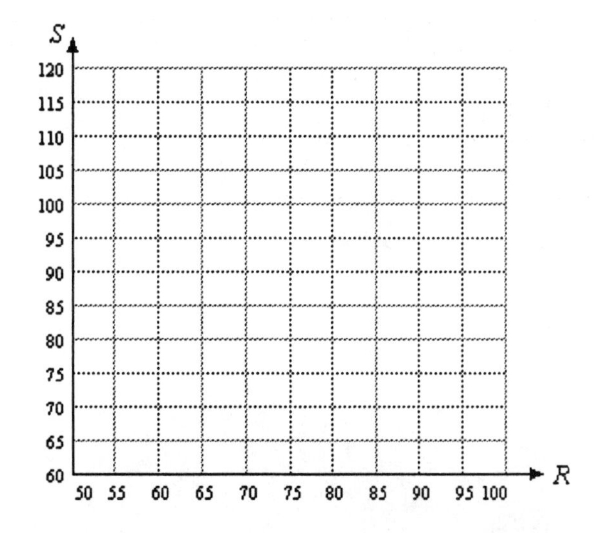

2.

Regular price	18	26	54	76	90	r
Calculation						
Delbert pays						

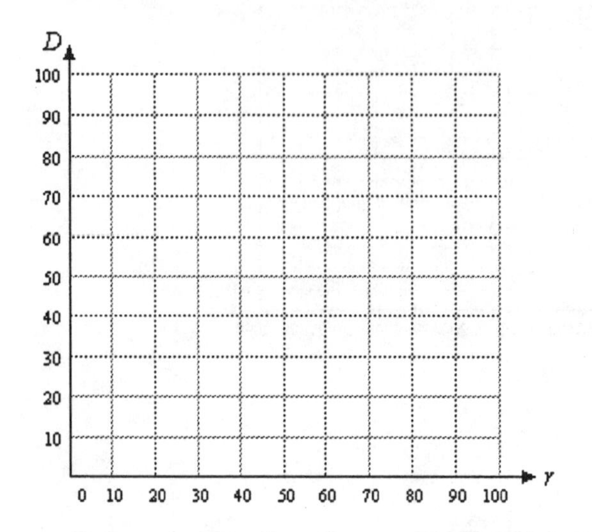

3.

Miles driven	40	60	95	120	145	170	d
Calculation							
Miles remaining							

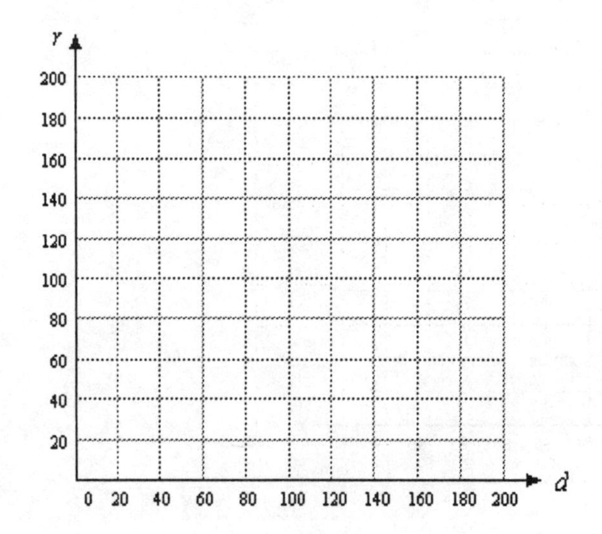

4.

Hours traveled	2	3	5	8	10	h
Calculation						
Miles traveled						

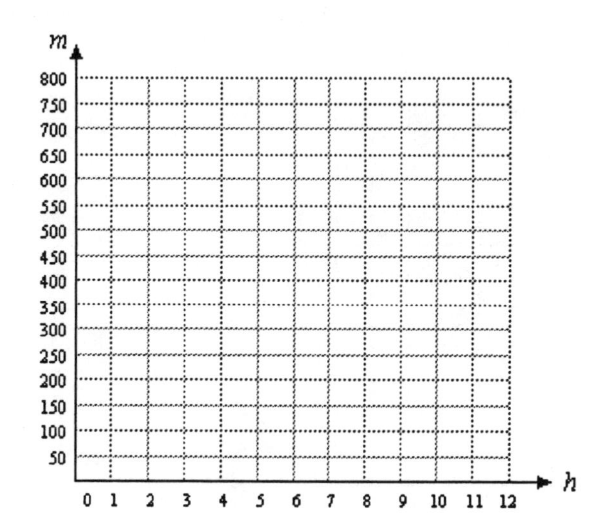

5.

Total bill	24	30	33	45	54	81	b
Calculation							
Milton's share							

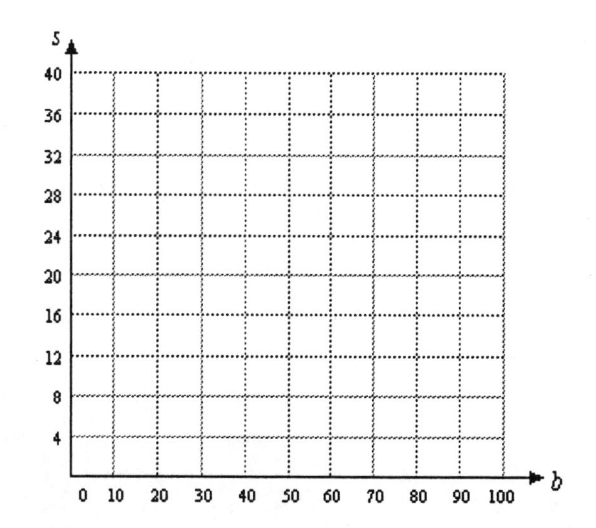

6.

Bill	20	24	36	44	50	55	*b*
Calculation							
Tip							

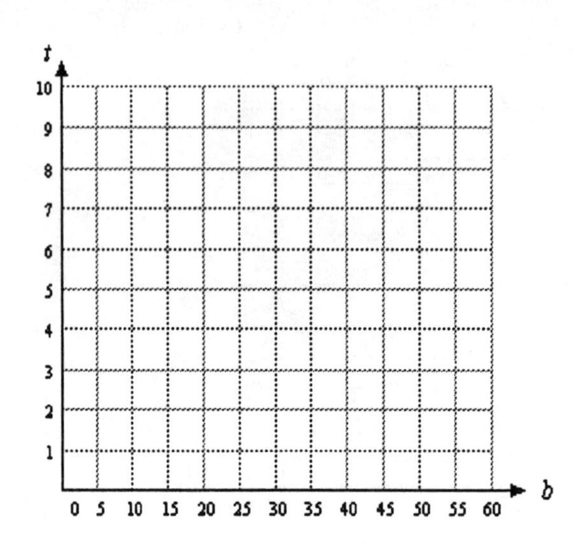

Section 1.4 Algebraic Expressions

Exercise 1

a. Ten more than the number of students

Steps 1-2

Step 3

b. Five times the height of the triangle

Steps 1-2

Step 3

c. 4% of the original price

Steps 1-2

Step 3

d. Two and a quarter inches taller than last year's height.

Steps 1-2

Step 3

Exercise 2

a. Sixty dollars less than first-class airfare

Steps 1-2

Step 3

b. The quotient of the volume of the sphere and 6

Steps 1-2

Step 3

c. The ratio of the number of gallons of alcohol to 20

Steps 1-2

Step 3

d. The current population diminished by 50

Steps 1-2

Step 3

Exercise 3 Evaluate the algebraic expression in Example 6 to complete the table below showing the sale price for various appliances

p	120	200	380	480	520
s					

Homework 1.4

51.

x	0	5	15	20	25	30
y	15	20	30			

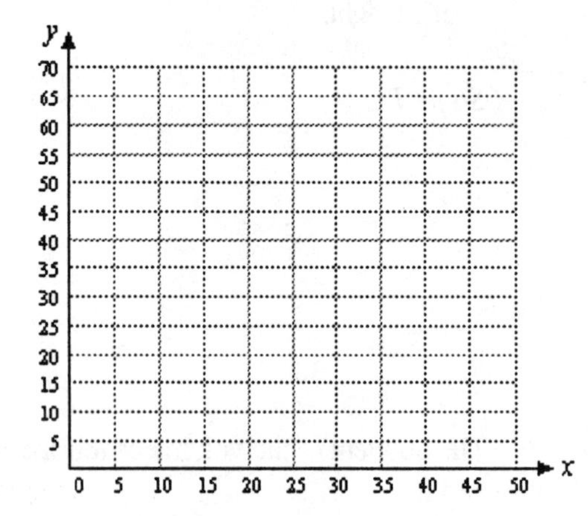

52.

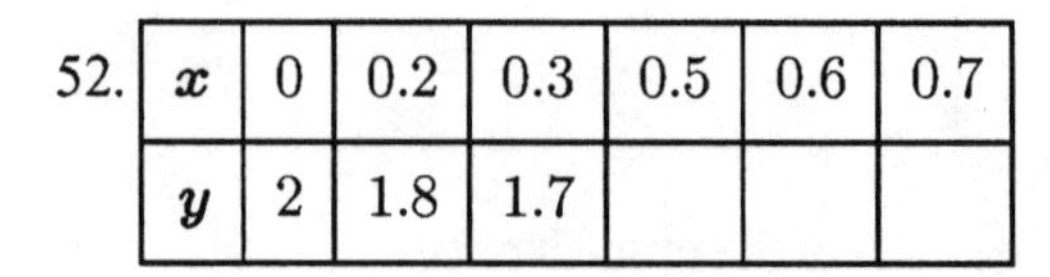

x	0	0.2	0.3	0.5	0.6	0.7
y	2	1.8	1.7			

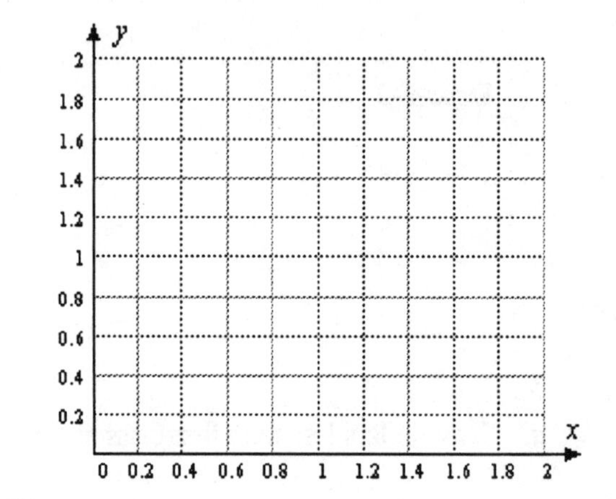

53.

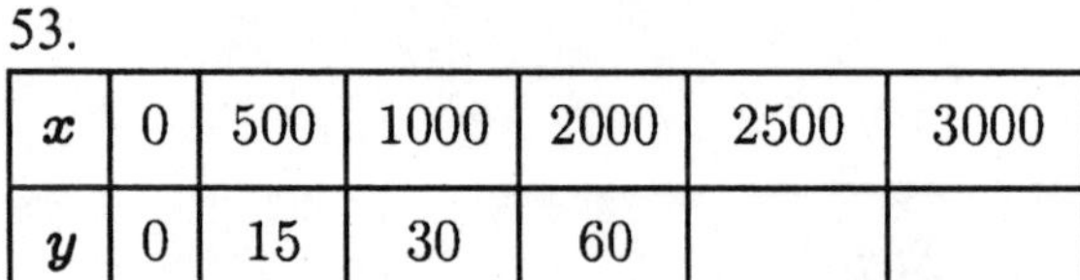

x	0	500	1000	2000	2500	3000
y	0	15	30	60		

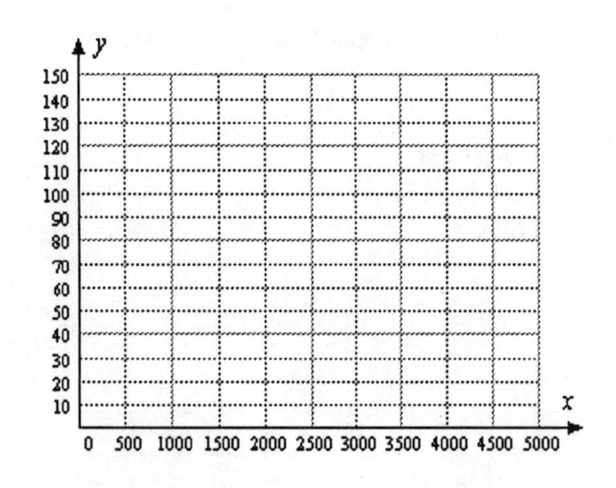

54.

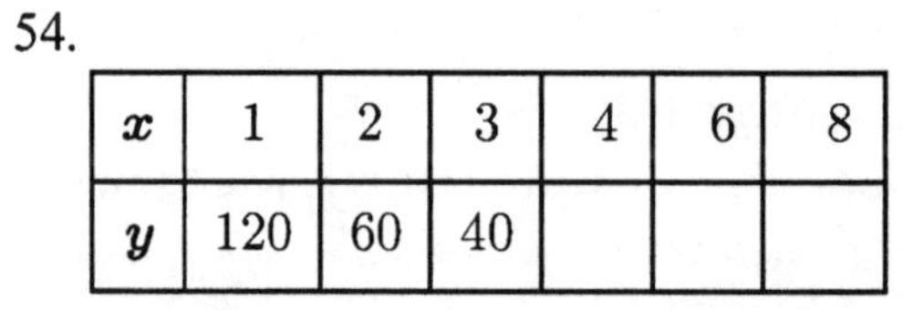

x	1	2	3	4	6	8
y	120	60	40			

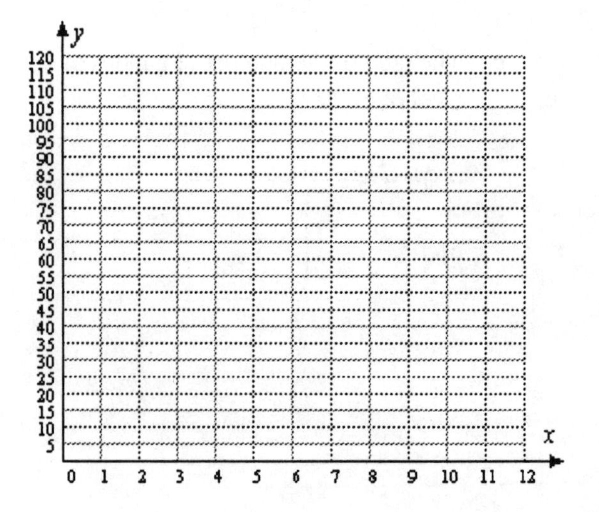

Section 1.5 Graphs of Equations

Exercise 1 Locate on the graph each of the ordered pairs listed in Table 1.17, and make a dot there. Label each point with its coordinates.

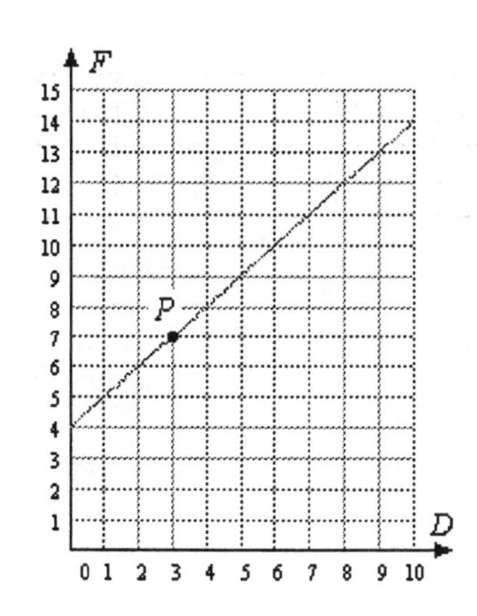

$F = D + 4$

D	F
0	4
2	6
3	**7**
4	8
6	10
9	13

Exercise 2 By substituting its coordinates into the equation, verify that each point you labeled in Exercise 1 represents a solution of the equation $F = D + 4$. (Use the space next to the table.)

Exercise 3 The Harris Aircraft company gave all its employees a 5% raise.

a. Write an equation that gives each employee's raise, R, in terms of his or her present salary, S.

b. Graph your equation as follows:

Step 1 Complete the table of values.

S	18,000	24,000	32,000	36,000
R				

Step 2 Choose scales for the axes in the figure, and label the axes.

Step 3 Plot points and draw the graph.

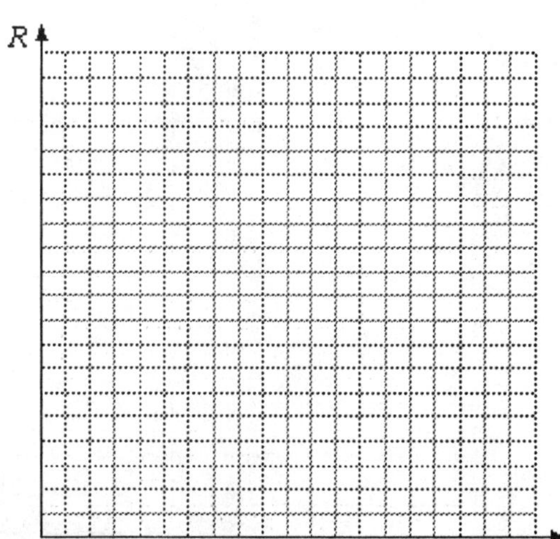

Midchapter Review

11.

Country	**Seats**	**Women**	**Percentage**
Canada	295	63	
France	577	63	
Great Britain	659	120	
Israel	120	9	
Japan	500	23	
Mexico	500	71	
South Africa	400	100	
Sweden	349	141	
United States	435	51	

16.

Interest	1150	1000	750	620	480
Calculation					
Principle					

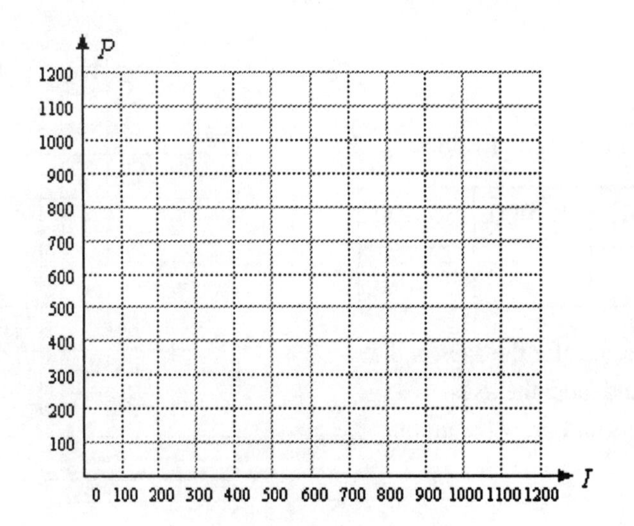

17.

Year	High School Graduates	College Graduates	Percent
1970	9567	13,264	
1975	13,542	17,477	
1980	19,469	24,311	
1985	23,853	32,822	
1990	26,653	39,238	

Section 1.6 Solving Equations

Exercise 1 Complete the table showing Liz's wages in terms of the number of hours she worked.

h	3	5	8	10
w				

Exercise 2a. Choose any number for x: $x =$ ________.

Multiply your number by 5: $5x =$ ________.

Divide the result by 5: $\frac{5x}{5} =$ ________.

b. Choose any number for x: $x =$ ________.

Add 4 to your number: $x + 4 =$ ________.

Subtract 4 from the result: $x + 4 - 4 =$ ________.

Exercise 3 One of Aunt Esther's Chocolate Dream cookies contains 42 calories, so if you eat d cookies you will gain c calories, where $c = 42d$. If Albert consumed 546 calories, how many cookies did he eat? Use trial and error to help you solve the equation $546 = 42d$.

d	5	6	7	8	9	10	11	12	13	14	15
c											

Answer:

Exercise 4 Fernando plans to share an apartment with three other students and split the rent equally.

a. Let r stand for the rent on the apartment and s for Fernando's share. Write an equation for s in terms of r.

b. Fill in the table.

r	260	300		360		480
s			80		105	

c. Explain how you found the unknown values of s.

Explain how you found the unknown values of r.

Exercises:

5. $5 + y = 9$

6. $x - 4 = 12$

7. $6z = 24$

8. $\frac{w}{3} = 6$

Exercise 9 The distance from Los Angeles to San Francisco is approximately 420 miles. How long will it take a car traveling at 60 miles per hour to go from Los Angeles to San Francisco?

Answer First decide which formula is appropriate.

formula: ______________________ .

Which of the variables in the formula is unknown? _______________ .

What values are given for the other variables in the formula? ____________________ .

Substitute the known values into the formula:

Solve the equation for the unknown variable:

Answer:

Section 1.7 Problem Solving with Algebra

Exercise 1 A two-bedroom house costs $20,000 more than a one-bedroom house in the same neighborhood. The two-bedroom house costs $105,000. How much does the one-bedroom house cost?

Solution ***Step 1*** Choose a variable for the unknown quantity.

Cost of the one-bedroom house: __________ .

Step 2 Write an equation in terms of your variable.

__________ + __________ = __________
cost of one-bedroom   cost of two-bedroom

Step 3 Solve your equation.

The one-bedroom house costs ______________ .

Exercise 2 A restaurant bill is divided equally by seven people. If each person paid $8.50, how much was the bill?

Solution ***Step 1*** Choose a variable for the unknown quantity. (What are we asked to find?)

Step 2 Write an equation. Express each person's share in two different ways.

Step 3 Solve your equation.

The bill was ______________ .

Exercise 3 Iris got a 6% raise. Her new salary is \$21 a week more than her old salary. What was her old salary?

Solution ***Step 1*** Choose a variable for the unknown quantity.

Step 2 Write an equation. Express Iris's raise in two different ways.

Step 3 Solve your equation.

Iris's old salary was ________________ .

Exercise 4 Figure 1.35 shows a graph of the equation from Example 1, $s = 0.35b$. Note that both axes of the graph are scaled in *thousands* of dollars. Use the graph to estimate the answers to the questions in Example 1, namely:

a. Evaluate the equation for $b = \$1{,}800{,}000$.

b. Solve the equation for $s = \$875{,}000$.

Indicate on the graph how you obtained your estimates.

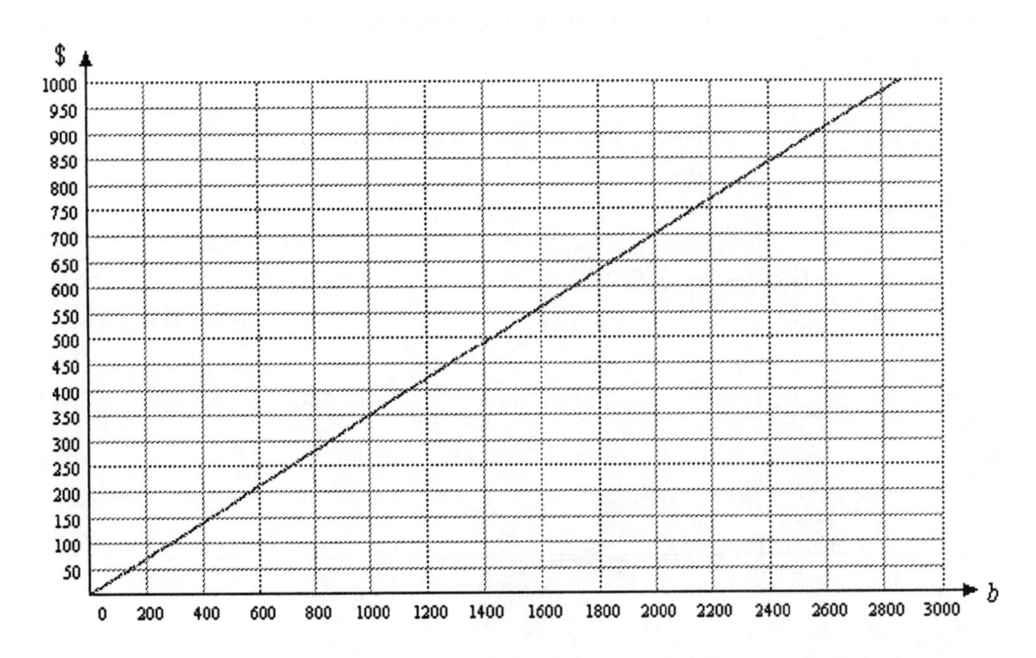

Section 1.8 Order of Operations

Exercise 1 Simplify each expression.

a. $30-17-5+4$

b. $72 \div 4 \cdot 3 \div 6$

Exercise 2 Simplify each expression. Perform multiplications before additions or subtractions.

a. $12-6\left(\frac{1}{2}\right)$

b. $2(3.5)+10(1.4)$

Exercise 3 Simplify $12+24 \div 4 \cdot 3+16-10-4$

Exercise 4 Simplify each expression.

a. $28-3(12-2 \cdot 4)$

b. $12+36 \div 4(9-2 \cdot 3)$

Exercise 5 Simplify $\frac{8-2(6-4)}{(8-2)6-4}$.

Exercise 6 Simplify $19+5\left[4(22-19)-\frac{12}{2}\right]$

Exercise 7a. Write the expression $\frac{16.2}{(2.4)(1.5)}$ in the on-line form.

b. Use a scientific calculator to simplify the expression.

Answers: a. ***b.***

Section 1.9 More Algebraic Expressions

Exercise 1 Write algebraic expressions for the following phrases.

a. The sum of 1 and 3 times t, divided by 2

b. The ratio of 25 to the sum of p and 6

2. Which expression represents "6 times the sum of x and 5"?

a. $6x+5$ *b.* $6(x+5)$

3. Which expression represents "$\frac{1}{2}$ the difference of p and q"?

a. $\frac{1}{2}(p-q)$ *b.* $\frac{1}{2}p-q$

4. Which expression represents "4 less than the product of 6 and w"?

a. $6w-4$ *b.* $4-6w$

5. Which expression represents "2 less than the quotient of 10 and z"?

a. $\frac{10}{z}-2$ *b.* $2-\frac{10}{z}$

6. Explain why $12x \div 3y$ is not the same as $\frac{12x}{3y}$

Exercise 7 Neda decides to order some photo albums as gifts. Each album costs \$12, and the shipping cost is \$4.

a. What is Neda's bill if she orders 3 albums? 5 albums?

b. Describe in words how you calculated your answers to part (a).

c. Fill in the table below.

Number of albums	2	3	4	6	8	10	15
Calculation	$12(2)+4$						
Neda's bill	**28**						

d. Write an algebraic expression for Neda's bill if she orders a albums.

e. Write an equation that gives Neda's bill, B, in terms of a.

Exercise 8a. Emily bought five rose bushes for her garden. Each rose bush cost $9 plus tax. If the tax on one rose bush is t, write an expression for the total amount Emily paid.

b. Megan would like to buy a kayak on sale. She calculates that it costs $40 less than three weeks salary. If Megan makes w dollars per week, write an expression for the price of the kayak.

Exercises Fill in the tables to evaluate each expression in two steps. Note how the order of operations is different in parts (a) and (b).

9a. $8+3t$

t	$3t$	$8+3t$
0		
2		
7		

b. $3(t+8)$

t	$t+8$	$3(t+8)$
0		
2		
7		

10a. $6+\frac{x}{2}$

x	$\frac{x}{2}$	$6+\frac{x}{2}$
4		
8		
9		

b. $\frac{6+x}{2}$

x	$6+x$	$\frac{6+x}{2}$
4		
8		
9		

Exercise 11 Evaluate $8(x+xy)$ for $x=\frac{1}{2}$ and $y=6$.

Homework 1.9

1.

Number of weeks	2	4	5	6	10	15	20
Calculation	$5000-200(\mathbf{2})$						
Savings left	**4600**						

2.

Number of units	3	5	8	9	12	15	16
Calculation	$50+15(\mathbf{3})$						
Tuition	**95**						

3.

Income	5000	7000	12,000	15,000	20,000	24,000	30,000
Calculation	$0.12(\mathbf{5000}-2000)$						
State Tax	**360**						

4.

Car wash proceeds	100	200	500	600	800	1000	1100
Calculation	$(900+\mathbf{100})\div 4$						
Each charity's share	**250**						

5.

z	$5z$	$5z-3$
2		
4		
5		

6.

m	$m-3$	$\frac{m-3}{4}$
6		
7		
9		

7.

Q	$12+Q$	$2(12+Q)$
0		
4		
8		

8.

w	$\frac{1}{w}$	$1-\frac{1}{w}$
1		
2		
3		

Section 1.10 Equations with Two or More Operations

Exercise 1

a. If you put on socks and then put on shoes, what operations are needed to reverse the process?

b. You leave home and bicycle north for 3 miles and then east for 2 miles. Suddenly you notice that you have dropped your wallet. How should you retrace your steps?

Exercise 2 Fill in the tables with the correct values.

a.

n	$3n$	$3n-5$
2		
5		
		7
		22

b.

m	$\frac{m}{4}$	$\frac{m}{4}+1$
8		
12		
		6
		2

Exercise 3 Consider the equation $3n-5=p$. Refer to Exercise 2a to help you answer the questions:

a. Let $n=2$. Explain how to find p in two steps.

b. Let $p=7$. Explain how to find n in two steps.

Exercise 4 Consider the equation $\dfrac{m}{4}+1=h$. Refer to Exercise 2b to help you answer the questions:

a. Let $m=8$. Explain how to find h in two steps.

b. Let $h=6$. Explain how to find m in two steps.

Exercises

5. Solve $3x+2=17$

6. Solve $\frac{4a}{5}=12$

7. Solve $\frac{b+3}{4}=6$

Homework 1.10

1.

x	$2x$	$2x+4$
3		
6		
		14
		20

2.

a	$3a$	$\frac{3a}{2}$
4		
8		
		9
		18

3.

q	$q-3$	$5(q-3)$
3		
		10
4		
		20

4.

w	$w+3$	$\frac{w+3}{2}$
7		
		4
1		
		2

35. Lori's history test had one essay question and some short-answer questions. Lori scored 20 points on the essay, and she gets 4 points for each correct short answer. Write and graph an equation for Lori's score, s, if she gives x correct short answers. If Lori's score is 76, how many correct answers did she give?

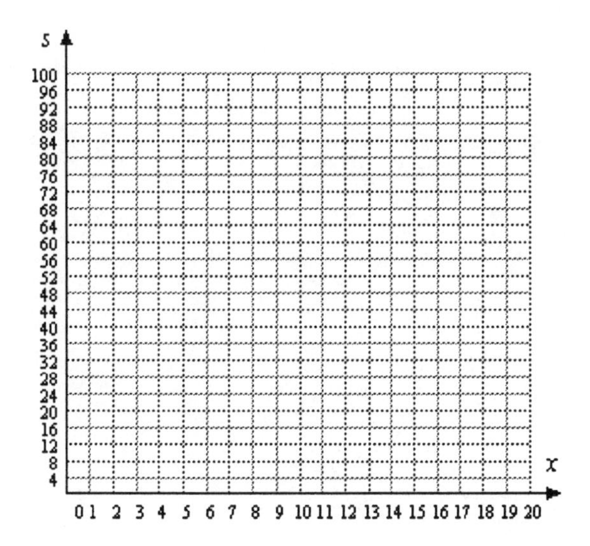

36. Yakov has saved \$300 and plans to add \$20 per week to his savings. Write and graph an equation for Yakov's savings, S, after w weeks. How long will it take Yakov's savings to grow to \$560?

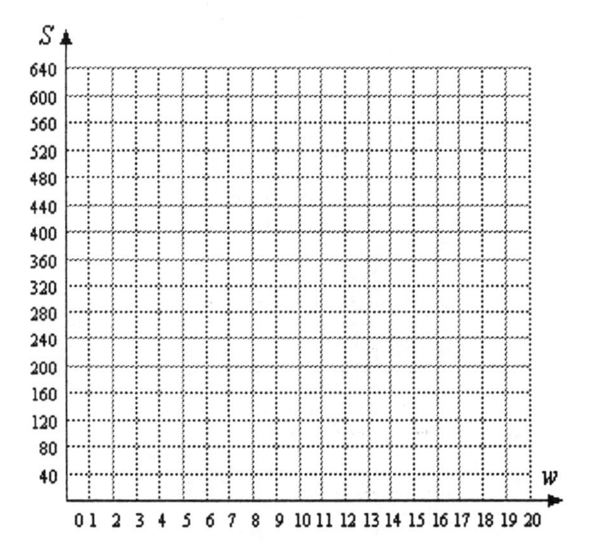

37. Gary employs 3 part-time waiters in his restaurant, all earning \$200 per week. He plans to give them all a raise. Write and graph an equation for Gary's new weekly payroll, P, if he raises each salary by r dollars. If Gary can afford \$654 a week for salaries, what raise can he give each waiter?

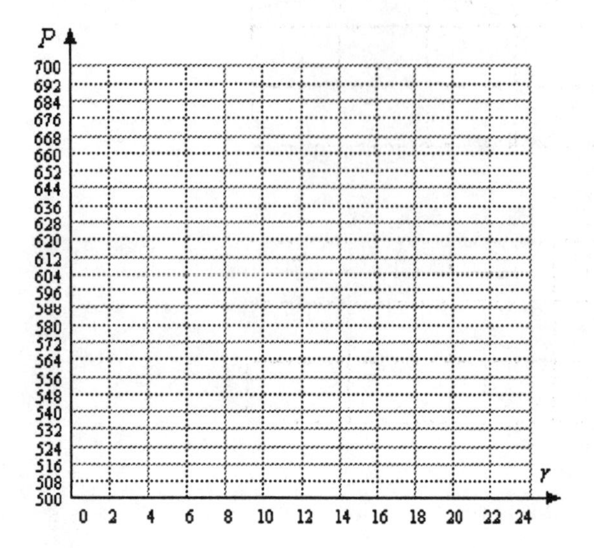

38. Heidi planted a live 6-foot Christmas tree, and it is growing 4.5 inches per year. Write and graph an equation for the height, h, of the tree after y years. (*Hint:* How many inches are in 6 feet?) How long will it take the tree to grow to 9 feet tall?

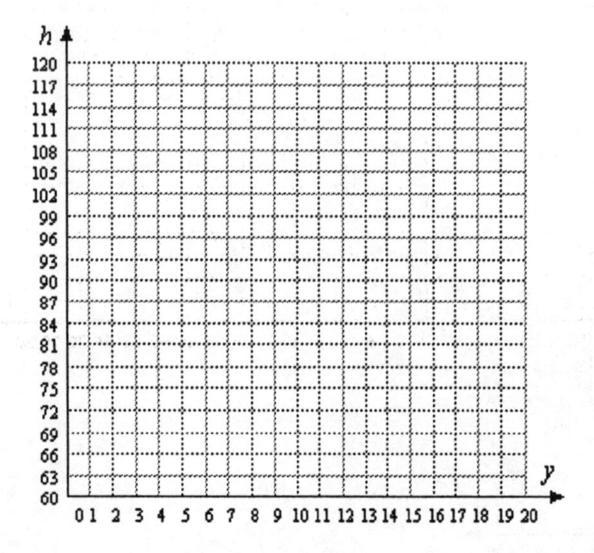

Chapter 1 Summary and Review

13.

Gallons of gas	3	5	8	10	11	12
Calculation						
Miles driven						

14.

Pages read	20	50	85	110	135	180
Calculation						
Pages left						

57.

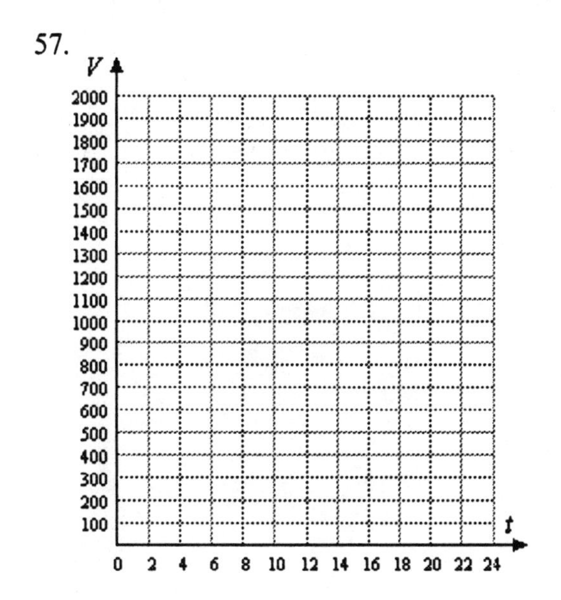

58.

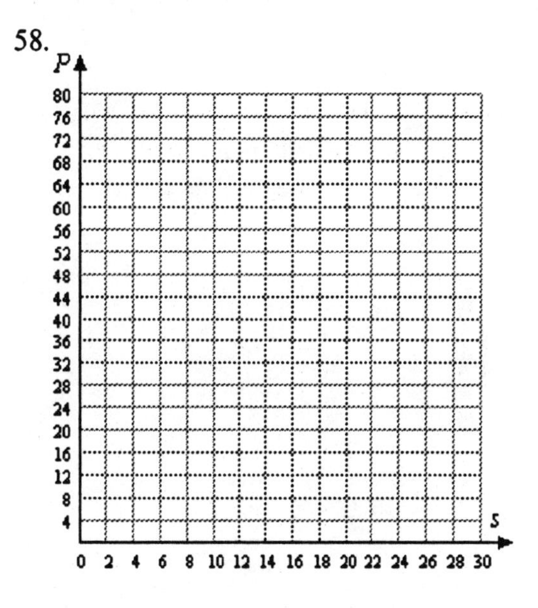

Chapter 2 Linear Equations

Section 2.1 Adding Signed Numbers

Exercise 1 Simplify each expression.

a. $|6| - |-3|$ *b.* $|8 - 3| - 2$

Answers:

Exercise 2 Replace the comma in each pair by the proper symbol: $<$ or $>$.

a. $1, -3$ *b.* $-9, -6$

Answers: *a.* *b.*

Exercise 3 *a.* List three solutions of the inequality $x > -4$.

b. Graph all the solutions on a number line.

Answers: *a.*

b.

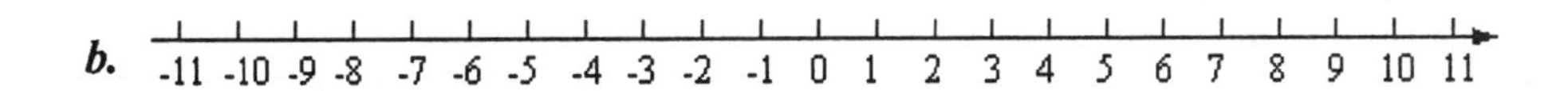

Exercise 4

a. $2 + 4 =$

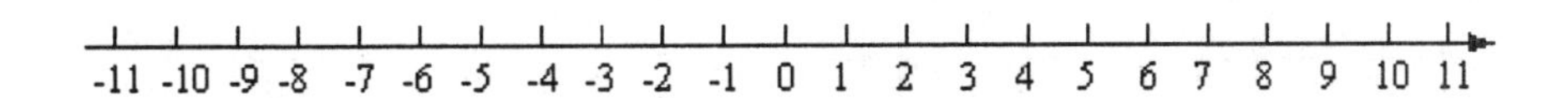

b. $(-4) + (-7) =$

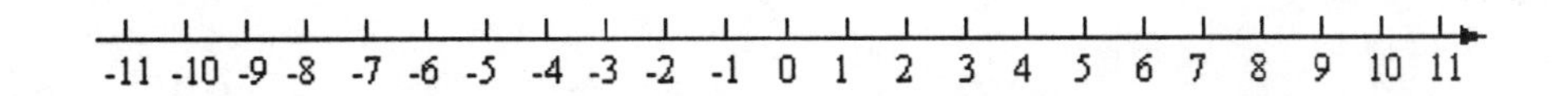

c. $(-6)+(-3)=$

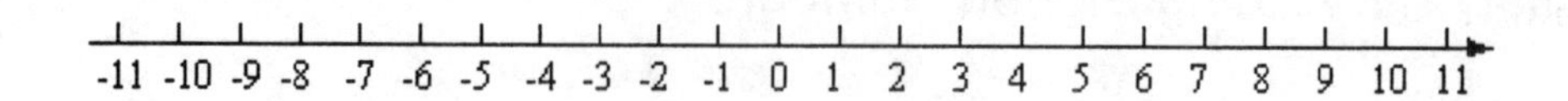

Exercise 5 Find the following sums without using a number line.

a. $(-9)+(-9)$ *b.* $(-14)+(-11)$

Answers: ***a.*** ***b.***

Exercise 6

a. $(+5)+(-3)=$

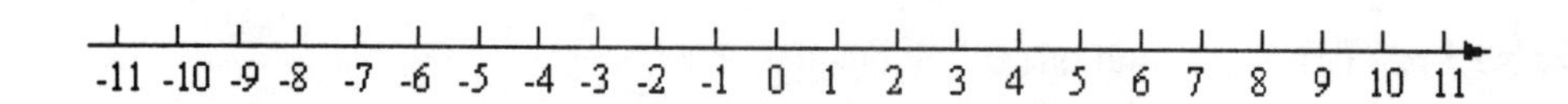

b. $(-7)+(+2)=$

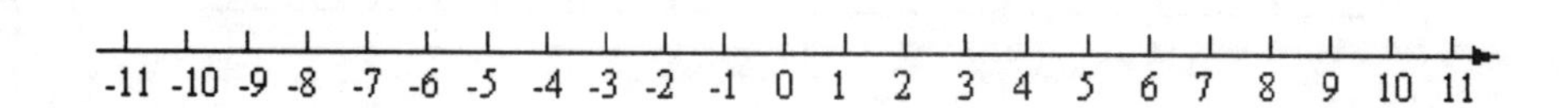

c. $(-5)+(+9)=$

-11 -10 -9 -8 -7 -6 -5 -4 -3 -2 -1 0 1 2 3 4 5 6 7 8 9 10 11

Exercise 7 Give a similar explanation for the answer to Exercise 6b.

Answer:

Exercise 8 Compute the following sums without using number lines.

a. $4+(-12)$ *b.* $15+(-9)$

Answers: *a.* *b.*

Exercise 9 Use the rules for addition to find the following sums.

a. $-6+(-3)$ *b.* $-8+(+3)$ *c.* $-7+19$

d. $18+(-10)$ *e.* $5+(-5)$ *f.* $-5+(-5)$

Answers: *a.* *b.*

c. *d.*

e. *f.*

Section 2.2 Subtracting Signed Numbers

Exercises

1a. $3+(-10)=$

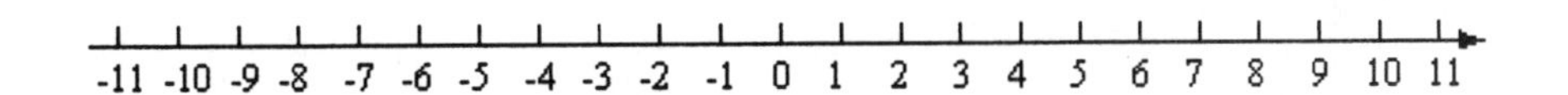

b. $3-(+10)=$

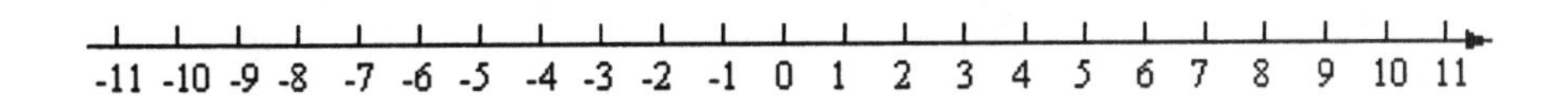

2a. $-6+(-3)=$

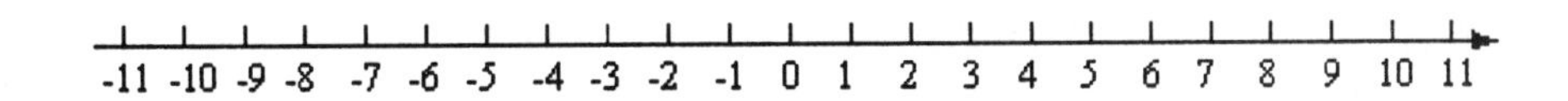

b. $-6-(+3)=$

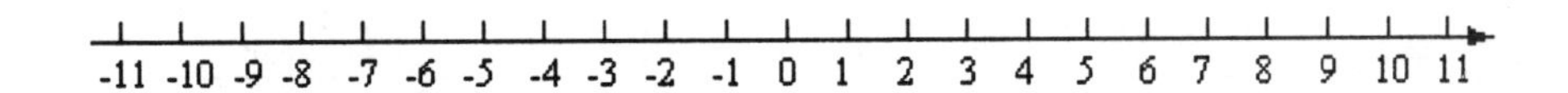

Exercise 3

a. $2-(-6)=$

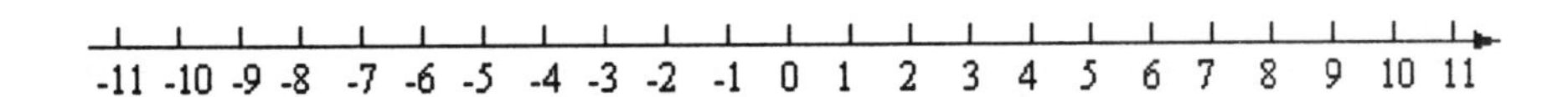

b. $-7-(-4)=$

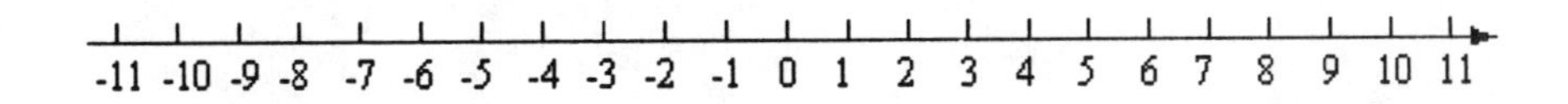

Exercise 4 Rewrite each subtraction problem as an addition, then compute the answer.

a. $3-(-9)$ *b.* $-4-(-7)$

Answers: a. *b.*

Exercise 5 Simplify $5-(+7)-3-(-2)$.

Answer:

Homework 2.2

75. The 10 students in a small class receive the following scores on a 10-point quiz:

$$4,\ 5,\ 6,\ 7,\ 7,\ 8,\ 8,\ 9,\ 10,\ 10$$

a. Compute the mean quiz score, denoted by $\overline{x}$. (See Appendix A.11 to review the definition of *mean*.)

b. For each quiz score, compute the *deviation from the mean*, $x-\overline{x}$, and fill in the table.

Score, x	Deviation from the mean, $x-\overline{x}$		Score, x	Deviation from the mean, $x-\overline{x}$
4	$4-\overline{x}=$		8	
5			8	
6			9	
7			10	
7			10	

c. Compute the average of the deviations from the mean. Can you explain why your answer makes sense?

76. Repeat Problem 75 for the following scores on a 20-point quiz:

$$20,\ 19,\ 17,\ 17,\ 16,\ 15,\ 15,\ 14,\ 12,\ 12,\ 10,\ 7$$

a. Compute the mean quiz score, denoted by $\overline{x}$. (See Appendix A.11 to review the definition of *mean*.)

b. For each quiz score, compute the *deviation from the mean*, $x - \overline{x}$, and fill in the table.

Score, x	Deviation from the mean, $x - \overline{x}$		Score, x	Deviation from the mean, $x - \overline{x}$
20	$20 - \overline{x} =$		15	
19			14	
17			12	
17			12	
16			10	
15			7	

c. Compute the average of the deviations from the mean. Can you explain why your answer makes sense?

Section 2.3 Multiplying and Dividing Signed Numbers

Exercise 1 Use the relationship between products and quotients to complete each statement. No calculation is necessary!

a. $\frac{8190}{26} =$ _____ because $26 \cdot 315 = 8190$

b. $62 \cdot$ _____ $= 83.7$ because $\frac{83.7}{62} = 1.35$

Exercise 2 Rewrite each division problem as a multiplication problem.

a. $\frac{144}{64} = 2.25$ *b.* $36 \div \frac{3}{8} = 96$

Answers: *a.* *b.*

Exercise 3 Find the quotient, if it exists.

a. $\frac{0}{18}$ *b.* $\frac{13}{0}$ *c.* $6 \div 0$ *d.* $0 \div 8$

Answers: *a.* *b.* *c.* *d.*

Exercise 4 Write the product $5(-4)$ as a repeated addition, and compute the product.

Answer:

The product of a positive number and a negative number is a __________ *number.*

The product of two negative numbers is a __________ *number.*

Exercise 5 Compute the following products.

a. $4(-2)$ *b.* $(-3)(-3)$ *c.* $-5(3)$

Answers: *a.* *b.* *c.*

Exercise 6 Determine each quotient by rewriting each division as an equivalent multiplication problem.

a. $\frac{6}{3} = \square$ because ______________________.

b. $\frac{6}{-3} = \square$ because ______________________.

c. $\frac{-6}{3} = \square$ because ______________________.

d. $\frac{-6}{-3} = \square$ because ______________________.

Exercise 7 Compute the following quotients.

a. $\frac{-25}{-5}$ *b.* $\frac{32}{-8}$ *c.* $\frac{-27}{9}$

Answers: *a.* *b.* *c.*

Exercise 8 Simplify each expression.

a. $-6(-2)(-5)$ *b.* $-6(-2)-5$

c. $-6(-2-5)$ *d.* $-6-(2-5)$

Answers: *a.* *b.*

c. *d.*

Section 2.4 Graphs of Linear Equations

Exercise 1 Give the coordinates of each point shown in Figure 2.16.

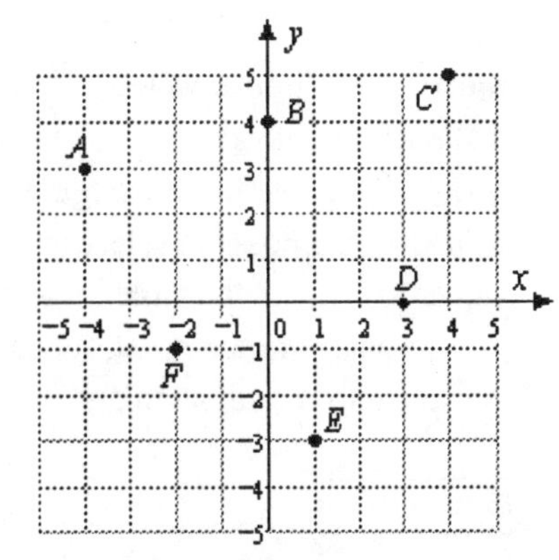

Figure 2.16

Answers:

A ***B*** ***C***

D ***E*** ***F***

Example 2 The temperature in Nome was $-12°$ at noon. It has been rising at a rate of $2°$ per hour all day.

a. Fill in the table below. T stands for the temperature, and h stands for the number of hours after noon. Negative values of h represent hours before noon.

b. Write an equation for the temperature, T, after h hours.

c. Graph your equation on the grid in Figure 2.17, using the values in the table.

Use your graph to answer the following questions:

d. What was the temperature 8 hours before noon?

e. When will the temperature reach $-4°$? When will the temperature reach $4°$?

f. How much did the temperature change from 2 pm to 6 pm?

h	T
-3	
-2	
-1	
0	
1	
2	
3	

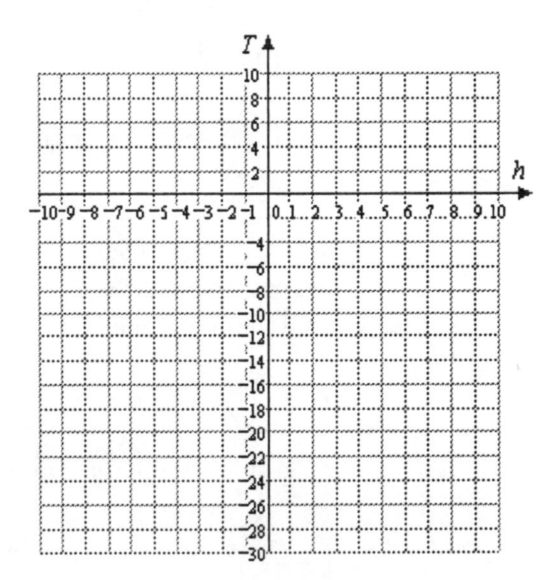

Figure 2.17

Example 3 Graph the equation $y = -2x + 6$.

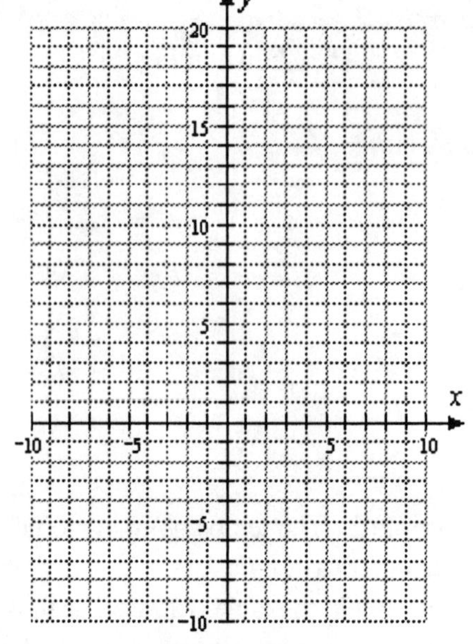

Figure 2.18

Solution **a.** Choose values for x and make a table of values. Make sure to choose both positive and negative x-values. (One possible selection of x-values appears in the table below.) To find the y-value for each point, substitute the x-value into the equation and evaluate.

x	y	
-3		$y = -2(\mathbf{-3}) + 6$
-1		$y = -2(\mathbf{-1}) + 6$
0		$y = -2(\mathbf{0}) + 6$
2		$y = -2(\mathbf{2}) + 6$
4		$y = -2(\mathbf{4}) + 6$

b. Look at the values in your table to help you choose scales for the axes. You will want to choose scales that include the largest and smallest values in your table. For this graph, we have chosen a scale from -10 to 20 on the y-axis.

c. Plot your points on the grid provided in Figure 2.18 and connect them with a smooth curve. All the points should lie on one straight line. □

Exercise 2 Use the graph in Figure 2.18 to answer the questions.

a. Find the value of $-2x + 6$ when $x = -5$.

b. Find the x-value for which $-2x + 6 = -10$.

Exercise 3 Which of the following points represent solutions to the equation whose graph is shown in Figure 2.19?

a. $(-3,-2)$ ***b.*** $(-6,-4)$

c. $(-4,0)$ ***d.*** $(3,-6)$

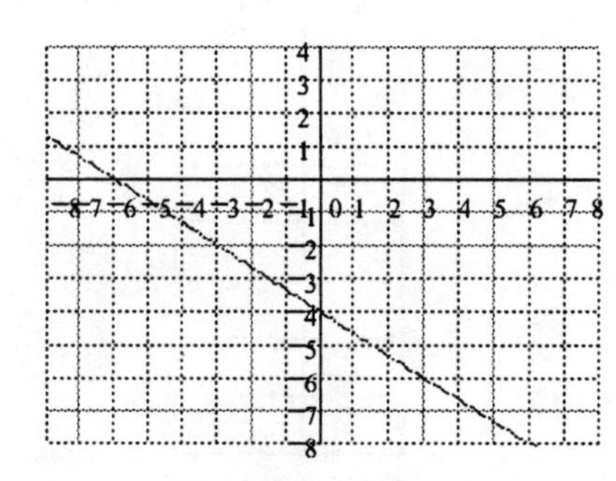

Figure 2.19

Exercise 4 Graph $y = \frac{-2}{3}x - 4$.

a. Fill in the table of values.

x	y
-9	
-3	
0	
3	
6	

$y = -\frac{2}{3}(-9) - 4$
$y = -\frac{2}{3}(-3) - 4$
$y = -\frac{2}{3}(0) - 4$
$y = -\frac{2}{3}(3) - 4$
$y = -\frac{2}{3}(6) - 4$

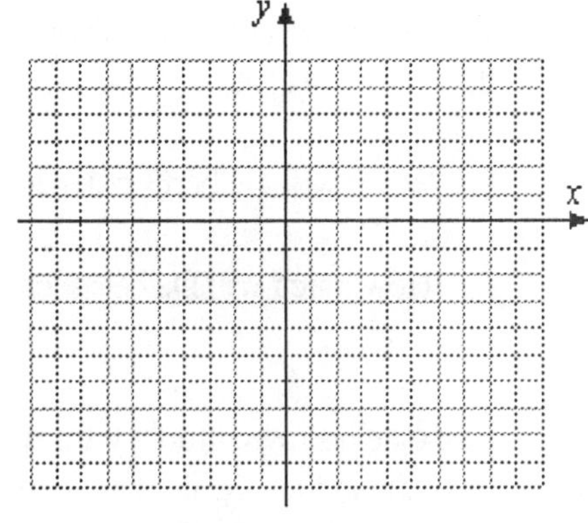

Figure 2.20

b. Choose appropriate scales and label the axes on the grid provided in Figure 2.20.
c. Plot your points and connect them with a straight line.

Homework 2.4

1. Delbert inherited \$5000 and has been spending money at the rate of \$100 per day. Right now he has \$2000 left.
   a. Fill in the table. B stands for Delbert's balance d days from now. Negative values of d represent days in the past.

d	-15	-5	0	5	15	20	25
B							

   b. Write an equation for Delbert's balance, B, after d days.

   c. Graph your equation on the grid, using the values in the table.

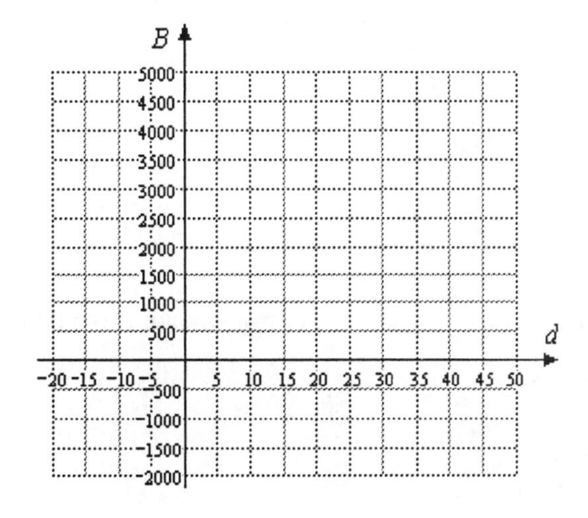

Use your graph to answer the following questions:

d. What was Delbert's balance 15 days ago?

e. When will Delbert's balance reach $500?

f. How much will Delbert spend from the beginning of day -5 to the beginning of day 40?

2. Francine borrowed money from her mother, and she currently owes her mother $750. She has been paying off the debt at a rate of $50 per month.

a. Fill in the table. F stands for Francine's financial status, and m is the number of months from now. Negative values of m represent months in the past. (Francine's current financial status is $-\$750$.)

m	-5	-2	0	2	6	10	12
F							

b. Write an equation for Francine's financial status, F, in terms of m.

c. Graph your equation on the grid, using the values in the table.

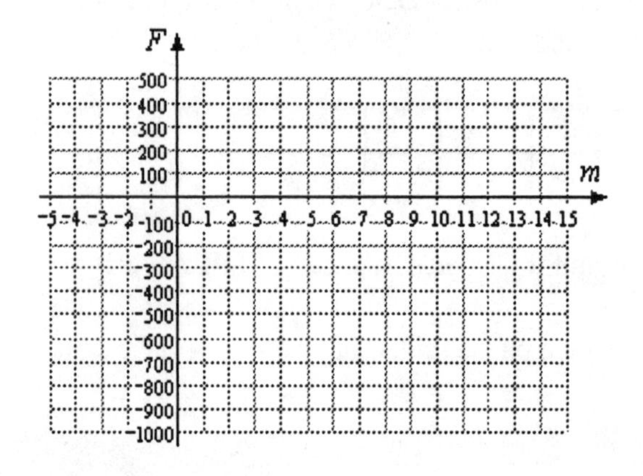

Use your graph to answer the questions:

d. What will Francine's financial status be 7 months from now?

e. When was Francine's financial status $-\$900$?

f. How much did Francine pay her mother from month 2 to month 9?

3. Jayme parks in the garage of her office building, 45 feet underground on Floor -5. Each floor of the building is 9 feet tall.
   a. Fill in the table. F stands for the floor, and E is the elevation on that floor. Negative elevations are below ground level.

F	-5	-2	0	1	2	4	6
E							

   b. Write an equation for the elevation, E, on floor F.

   c. Graph your equation on the grid, using the values in the table.

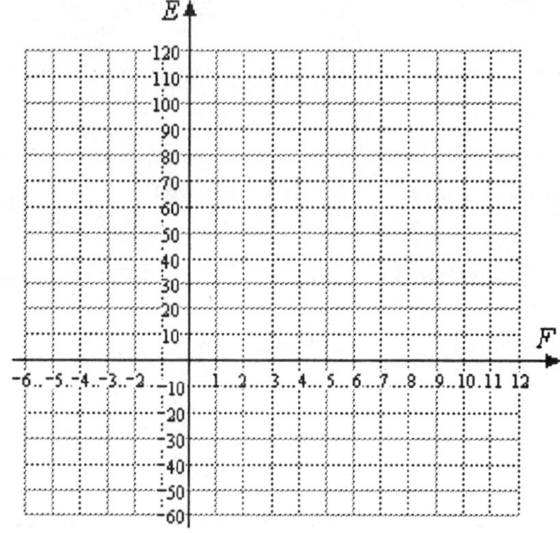

Use your graph to answer the questions:

   d. What is the elevation on floor -3?

   e. On what floor is the elevation 72 feet?

   f. What is Jayme's change in elevation when she takes the elevator from her parking level to her job on the tenth floor?

4. Ryan is cooling a chemical compound in the laboratory at a rate of 5 °F per hour. Right now the temperature of the compound is 35° F.
   a. Fill in the table. T stands for the temperature of the compound, and h is the number of hours from now.

h	-5	-3	-1	0	2	4	5
T							

   b. Write an equation for the temperature T of the compound after h hours.

c. Graph your equation on the grid, using the values in the table.

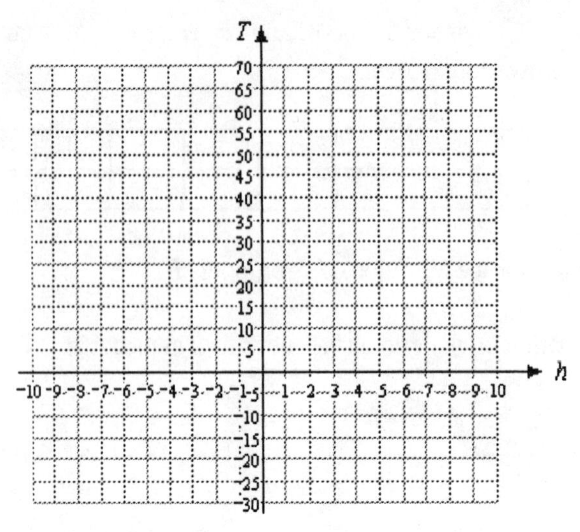

Use your graph to answer the questions:

d. When was the compound at room temperature, $70°$?

e. What will the temperature of the compound be 10 hours from now?

f. How much will the temperature drop between the third hour and the ninth hour?

☐ *Make a table of values and graph each equation on the grid provided. Extend your line far enough that it crosses both axes.*

5. $y = x + 3$

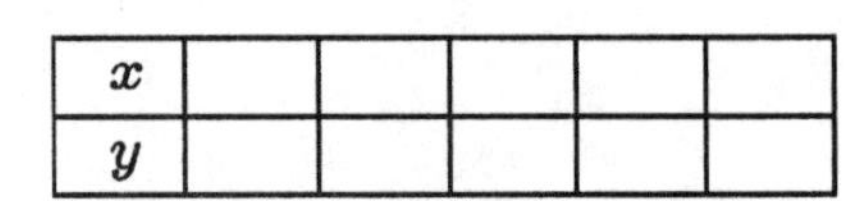

x					
y					

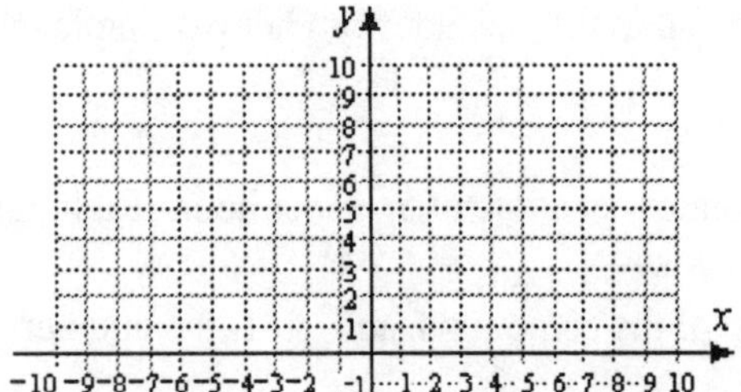
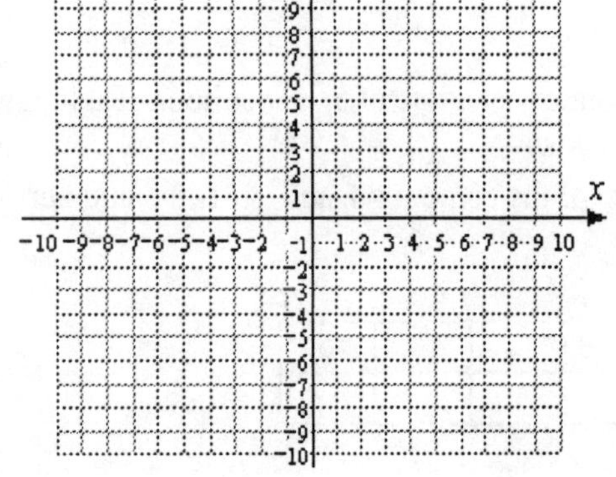

6. $y = -4 - x$

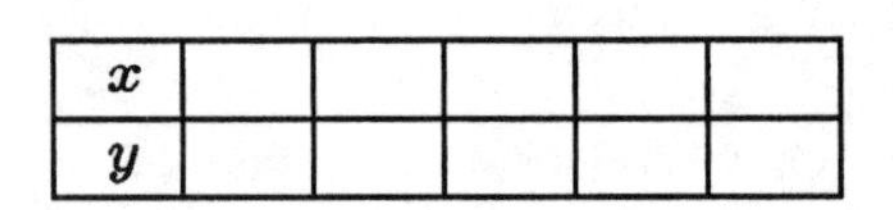

x					
y					

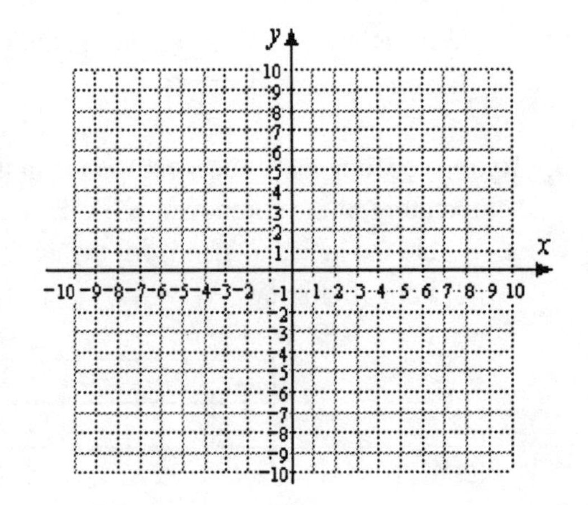

7. $y = 2x + 1$

x					
y					

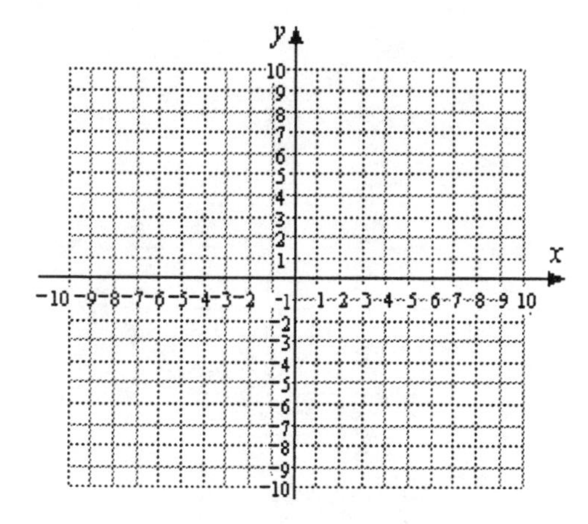

8. $y = 3x - 1$

x					
y					

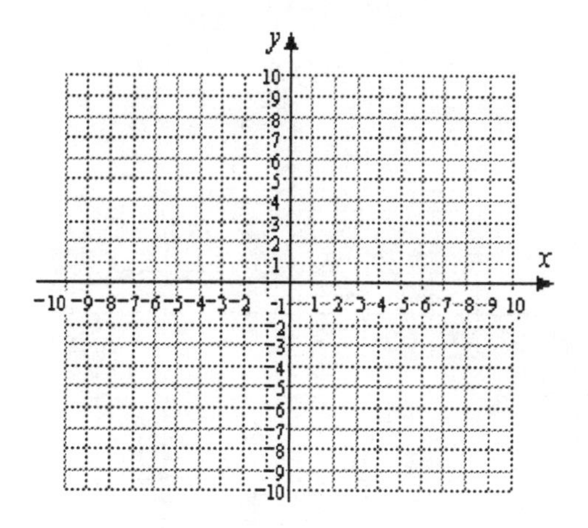

9. $y = -\frac{1}{2}x - 5$

x					
y					

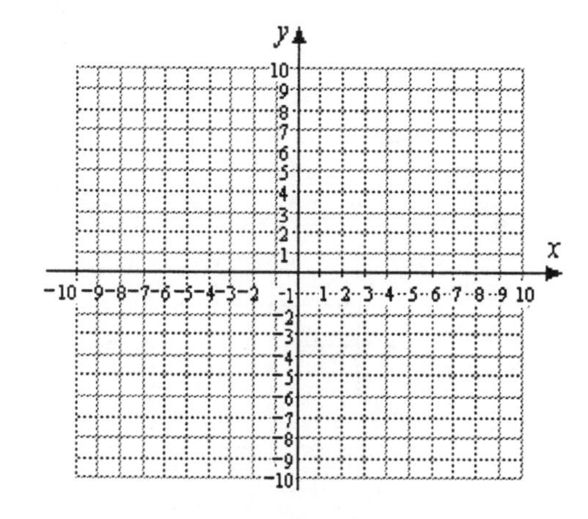

10. $y = \frac{3}{2}x + 2$

x					
y					

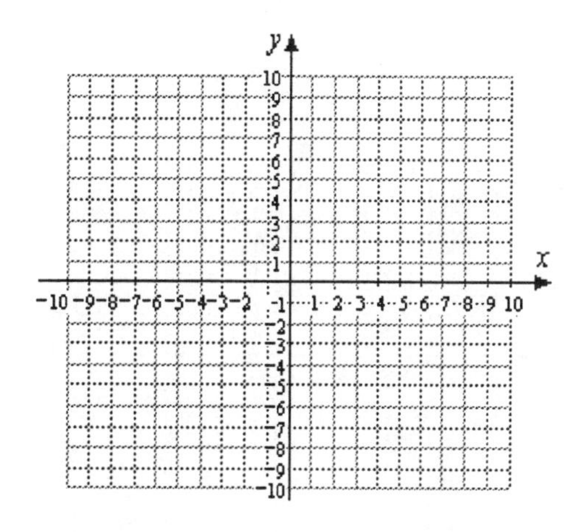

11. $y = \frac{5}{4}x - 4$

x					
y					

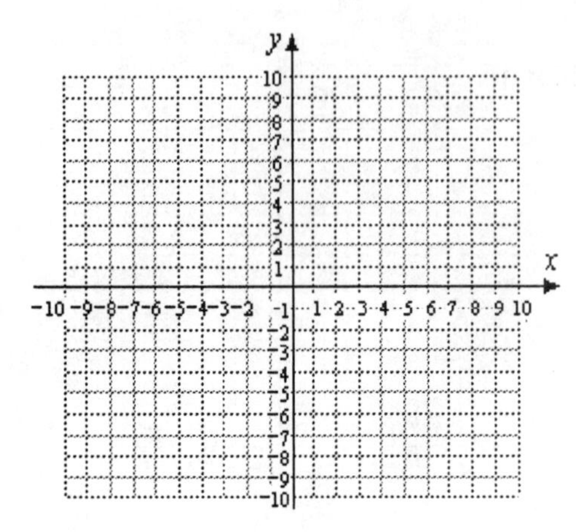

12. $y = \frac{-3}{4}x + 2$

x					
y					

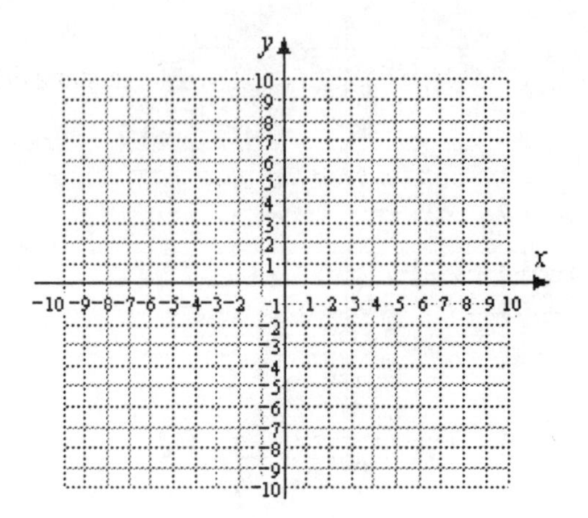

☐ *Plot each pair of points on the grid provided, then find the distance between the two points.*

23. a. $A(-6,8),\ B(-6,3)$
    b. $C(1,5),\ D(1,-7)$
    c. $E(7,-2),\ F(7,-8)$

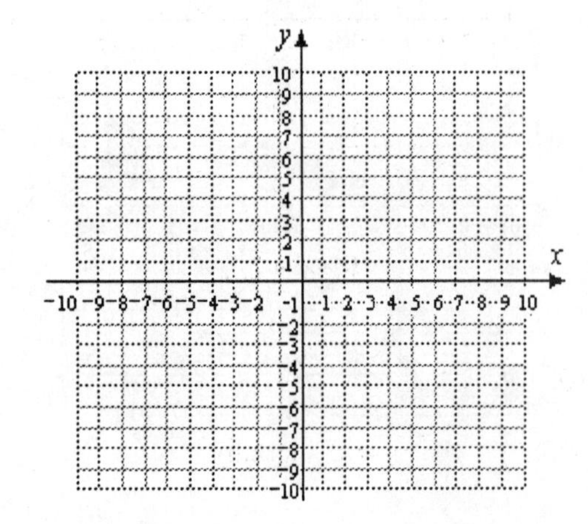

24. a. $P(-2,8),\ Q(5,8)$
    b. $R(-9,2),\ S(-3,2)$
    c. $T(1,-4),\ U(6,-4)$

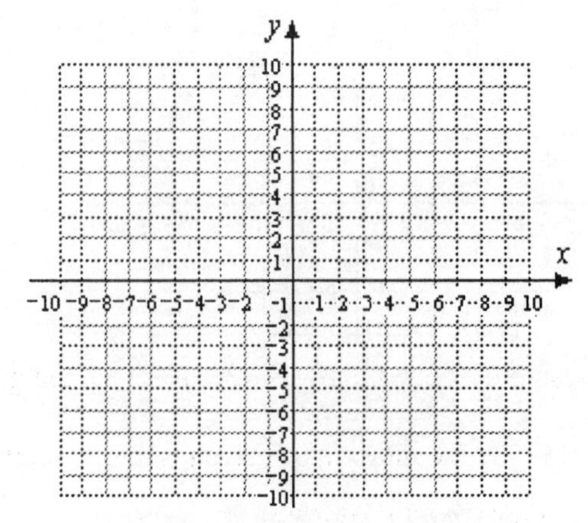

Section 2.5 Solving Linear Equations

Exercise 1 Solve the equation $-6 - \dfrac{2x}{3} = 8$

Step 1: Add 6 to both sides.

Step 2: Rewrite the fraction $-\dfrac{2x}{3}$ in standard form.

Step 3: Multiply both sides by 3.

Step 4: Divide both sides by -2.

Step 5: Check your solution.

Exercise 2 The point $(0, 8)$ also lies on the graph in Figure 2.23. This point gives us the solution to a certain equation in x. Write the equation and give its solution.

Answer:

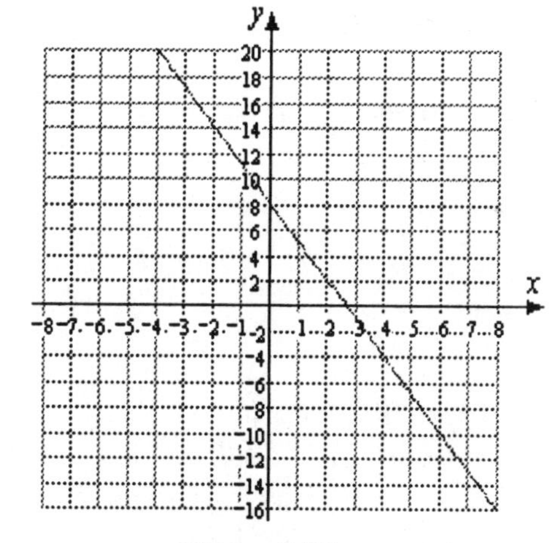

Figure 2.23

Exercise 3 The point $(-12,2)$ lies on the graph of $y = -6 - \dfrac{2x}{3}$.

a. Using the information given, solve the equation $-6 - \dfrac{2x}{3} = 2$ mentally.

b. Verify your solution algebraically.

Exercise 4 Use the graph in Figure 2.23 to solve the equation $8-3x=14$ as follows:

Step 1: Locate the point on the graph whose y-coordinate is 14.
Step 2: What is the x-coordinate of the point?
Step 3: Check that your x-value is a solution for $8-3x=14$.
Answer:

Exercise 5 Use algebra to find an exact solution for the equation in Example 3.

Answer:

Exercise 6a. Use the graph in Figure 2.24 to estimate the solution to the equation

$$9.6-2.4x=6.$$

b. Use the equation to check the accuracy of your approximation.

Answers: ***a.***
b.

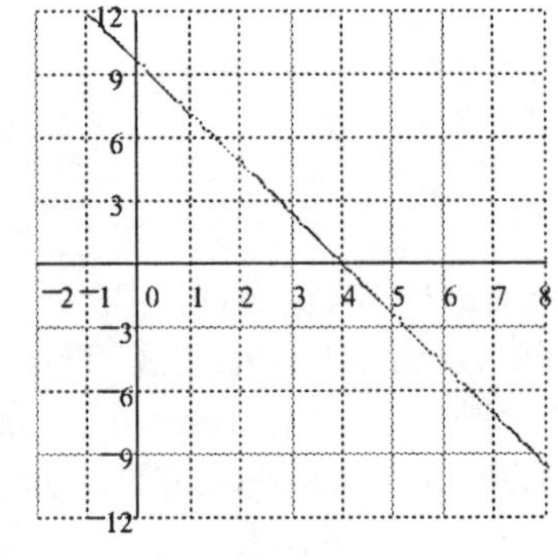

Figure 2.24

Homework 2.5

☐ *For part (c) of each problem, write an equation and solve it algebraically. Then use your graph to verify the solution.*

33. On a 100-point test, Lori loses 5 points for each wrong answer.
   a. Write an equation for Lori's score, s, if she gives x wrong answers.

   b. Complete the table of values and graph your equation on the grid.

x	2	5	6	12
s				

c. If Lori's score is 65, how many wrong answers did she give?

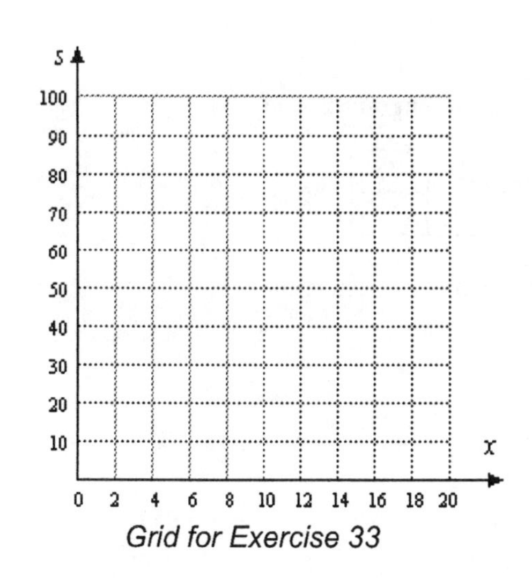

Grid for Exercise 33

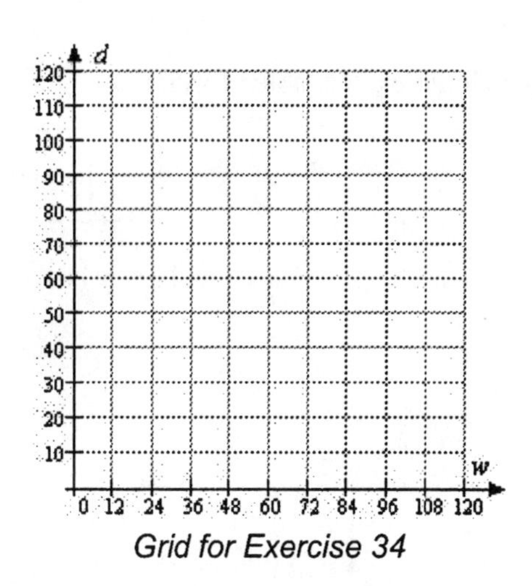

Grid for Exercise 34

34. The water in Silver Pond is 10 feet deep, but the water level is dropping at a rate of $\frac{1}{2}$ inch per week.
   a. Write an equation for the depth, d, of the pond after w weeks. (*Hint*: Convert all units to inches.)

   10 feet = ___________ inches

   $d =$

   b. Complete the table of values and graph your equation on the grid.

w	12	24	60	96
d				

   c. How long will it take until the depth of the pond is 8 feet? (*Hint*: Convert all the units to inches.)
   8 feet = ___________ inches

35. Larry bought a 10-pound box of laundry detergent, and every week he uses $\frac{1}{4}$ pound for the laundry.
    a. Write an equation for the amount of detergent, D, left after w weeks.

    b. Complete the table of values and graph your equation on the grid.

w	2	8	10	28
D				

    c. How long will it take until Larry has only $3\frac{1}{2}$ pounds of detergent left?

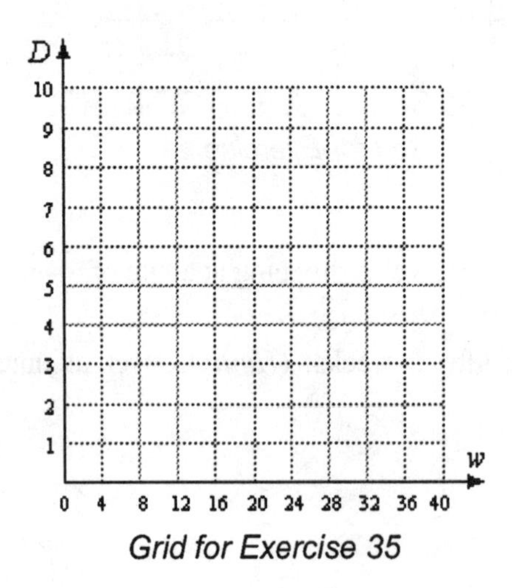

Grid for Exercise 35

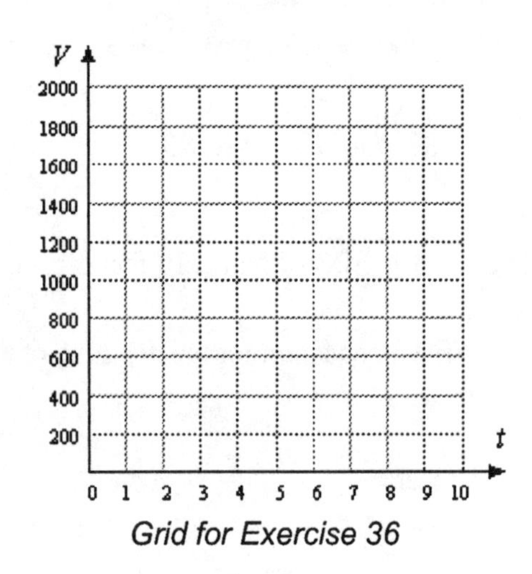

Grid for Exercise 36

36. A new computer workstation for a graphics design firm costs \$2000 and depreciates in value \$200 every year.
    a. Write an equation for the value, V, of the station after t years.

    b. Complete the table of values and graph your equation on the grid.

t	2	4	5	8
V				

    c. When will the station be worth only \$600?

Midchapter Review

23. The water level in the city reservoir was 30 feet below normal, so four days ago the city began diverting water from a nearby river. The level is rising at a rate of 2 feet per day.

   a. Fill in the table. W stands for the water level d days from today. Negative values of W represent feet below the normal water level.

d	-3	0	4	8	12
W					

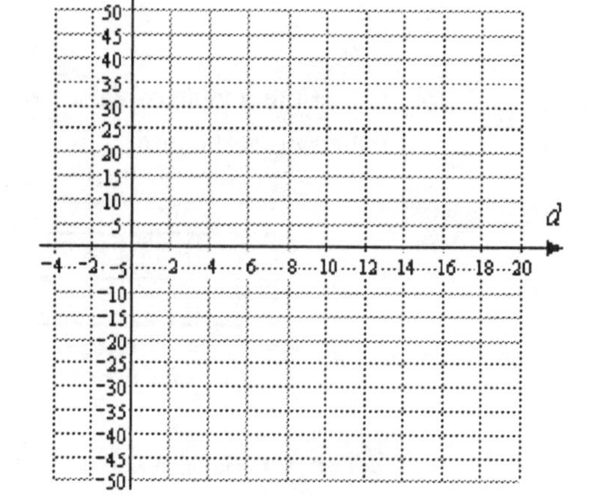

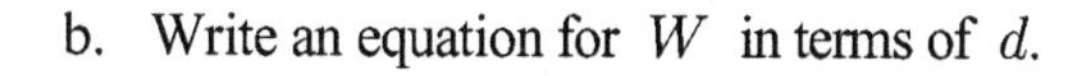

   b. Write an equation for W in terms of d.

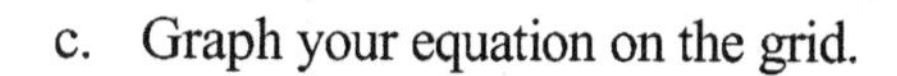

   c. Graph your equation on the grid.

   d. What will the water level be one week from now?

   e. When will the water reach its normal level?

   f. How much will the water level change from the sixth day to the fourteenth day?

34. Beryl is sailing in a hot air balloon at an altitude of 500 feet. She begins a slow descent at the rate of 15 feet per minute.

a. Write an equation for Beryl's altitude, h, after m minutes.

b. Complete a table of values and graph your equation on the grid.

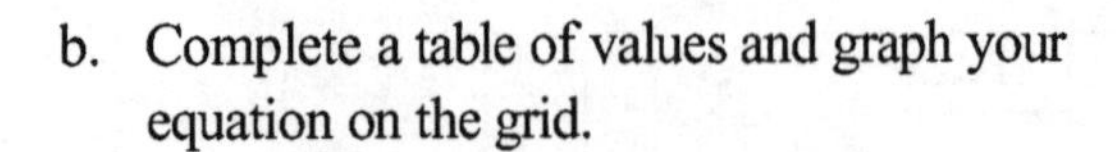

m				
h				

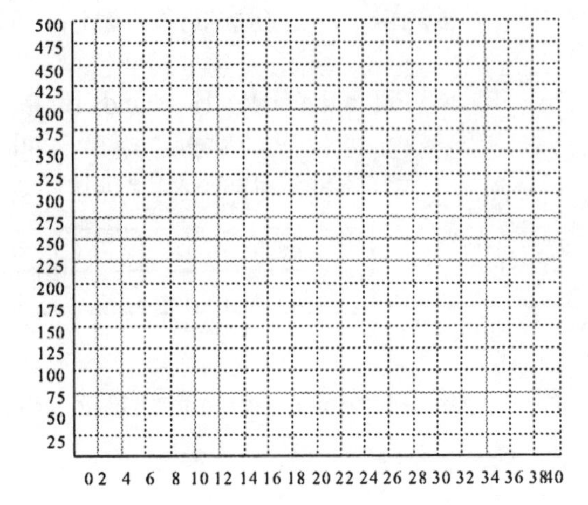

c. When will Beryl reach an altitude of 230 feet?

Write and solve an equation to answer this question, then verify the solution on your graph.

Section 2.6 Solving Linear Inequalities

Exercise 1 Fill in the correct symbol, $>$ or $<$, in each statement.

a. If $x > 8$, then $x - 7$____ 1.

b. If $x < -4$, then $3x$____ -12.

c. If $x > -2$, then $-9x$____ 18.

Exercise 2 Solve the inequality by using a graph:
$15 - 3x < 9$.

x	$15-3x$
-2	
-1	
0	
1	
2	
3	
4	
5	
6	
7	

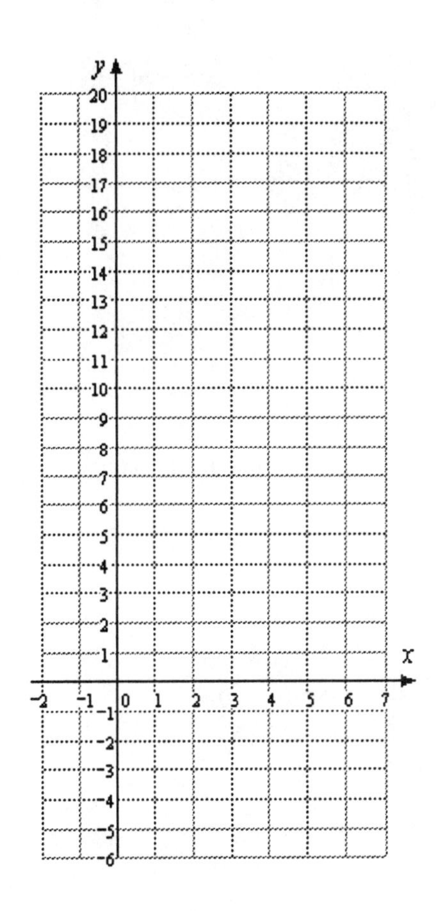

Exercise 3 Use one of the three rules stated above to solve each inequality. The first one is completed for you.

a.
$$\begin{array}{lcr} x - 8 & < & 3 \\ +8 & & +8 \\ \hline x & < & 11 \end{array}$$

What should you do to isolate x? **(Add 8 to both sides.)**

Should you reverse the direction of the inequality? **(No.)**

Is your answer reasonable:

Is 10 a solution? **$10 - 8 < 3$ is true, so 10 is a solution.**

Is 12 a solution? **$12 - 8 < 3$ is false, so 12 is not a solution.**

b. $\frac{x}{4} \geq -2$

What should you do to isolate x?
Should you reverse the direction of the inequality?

Is your answer reasonable:

Is -12 a solution?

Is -5 a solution?

c. $-5x > 20$

What should you do to isolate x?
Should you reverse the direction of the inequality?

Is your answer reasonable:

Is -10 a solution?

Is 2 a solution?

Exercise 4a. Solve $-8 < 4 - 3x < 10$.

b. Graph the solutions on the number line.

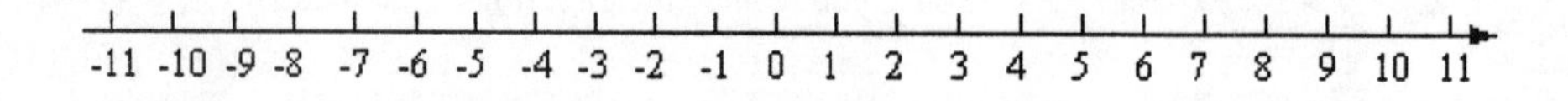

Homework 2.6

☐ *First, write and graph an equation relating the variables. Then, answer each question by solving an equation or inequality. Finally, use your graph to verify your answers.*

21. Today the high temperature was $56°$. If the temperature is cooling at $4°$ per day, write an equation for the temperature, T, after d days.

    Graph your equation on the grid.

    a. When will the temperature reach freezing $(32°)$?

    b. For which days will the temperature be below $-12°$?

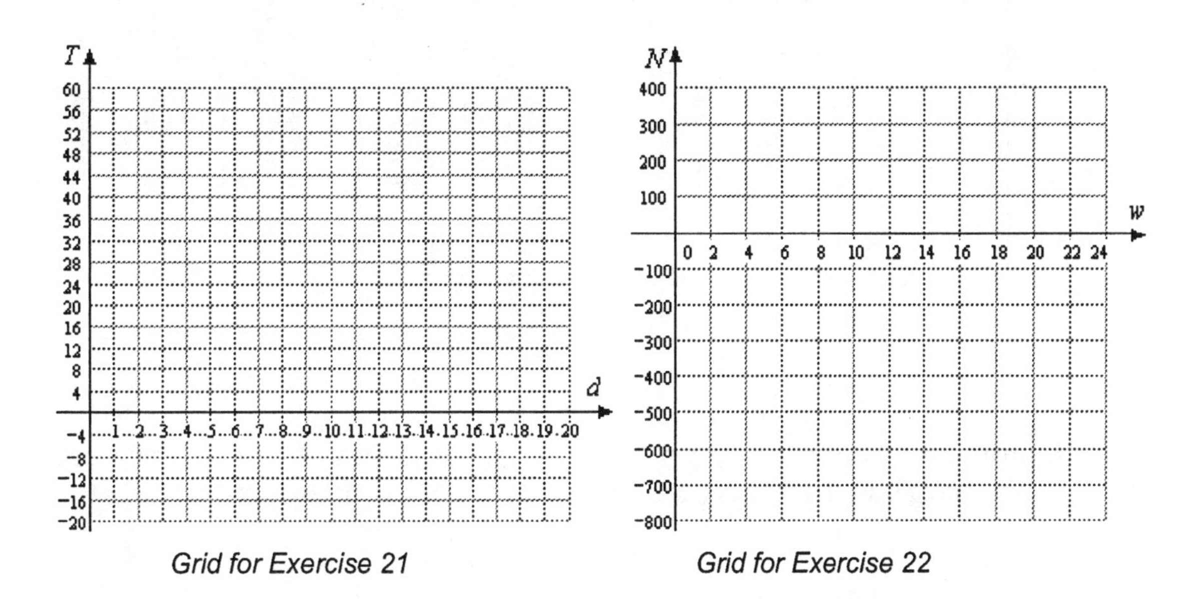

Grid for Exercise 21 *Grid for Exercise 22*

22. Yusuf is $\$750$ in debt, but he deposits $\$50$ a week into his savings. Write an equation for Yusuf's net worth, N, after w weeks.

    Graph your equation on the grid.

    a. When will Yusuf's net worth be $-\$100$?

    b. When will Yusuf's net worth be over $\$200$?

23. Francine is scuba diving at a depth of 200 feet, and is beginning to ascend at a rate of 15 feet per minute. Write an equation for Francine's elevation, h, after m minutes.

Graph your equation on the grid.

a. When will Francine's elevation be less than -20 feet?

b. When will Francine reach the surface?

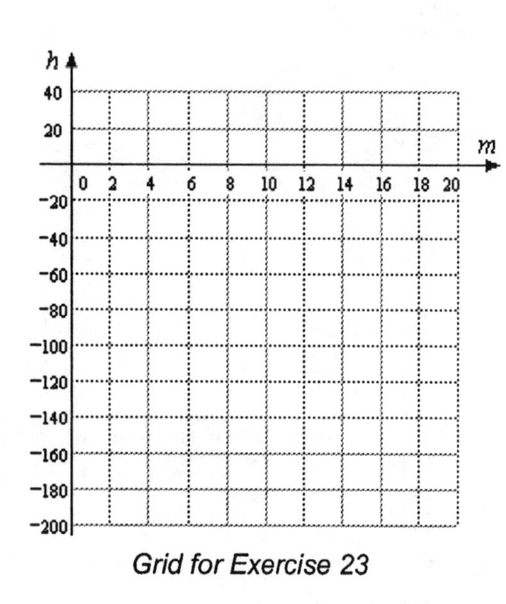

Grid for Exercise 23

Grid for Exercise 24

24. Briarwood School has a budget of $300,000, and it must spend $1500 on each student. Write an equation for the amount in Briarwood's account, B, if they enroll n students.

Graph your equation on the grid.

a. How many students are enrolled if Briarwood is in debt (i.e., it's account is overdrawn)?

b. How many students are enrolled if Briarwood's account is exactly $-\$18{,}000$?

Section 2.7 Intercepts of a Line

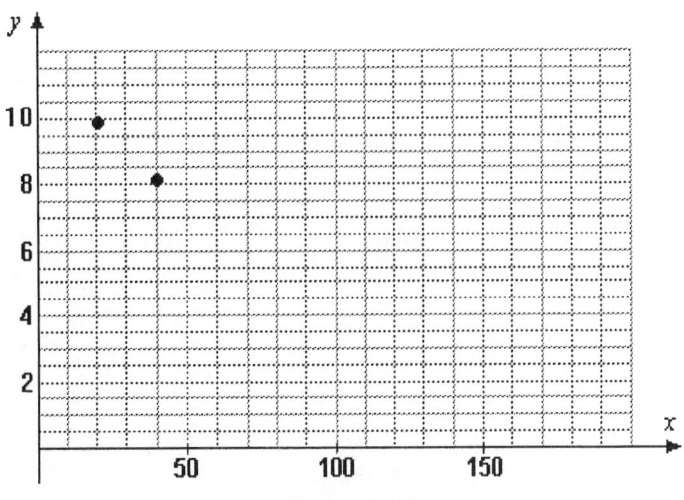

Figure 2.44

Exercise 1 a. How many acres of rainforest were present initially?

b. If we continue to clear the rainforest at the same rate, when will it be completely demolished?

Answers: a.

b.

Exercise 2 Find the x- and y-intercepts of the graph of $3x-2y=12$.

Answer:

Exercise 3 Graph the equation

$$2x=5y-10$$

by the intercept method.

x	y

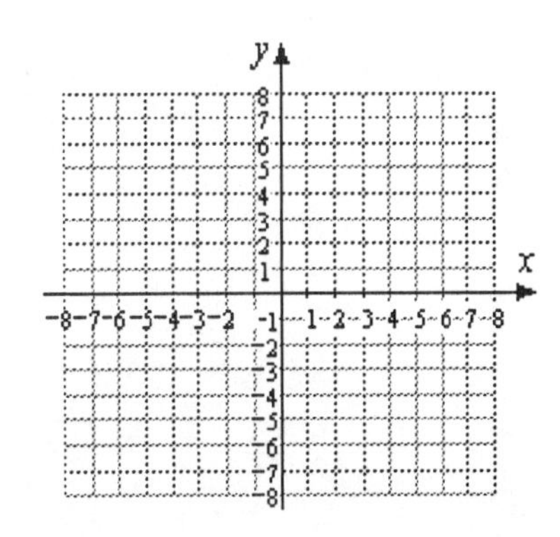

Section 2.8 Like Terms

Exercise 1a. Show that the expressions $6+2x$ and $8x$ are equal if $x=1$:

$6+2x=$ ________________,

$8x=$ ________________.

b. Show that the expressions $6+2x$ and $8x$ are *not* equal if $x=2$:

$6+2x=$ ________________,

$8x=$ ________________.

c. Are the expressions $6+2x$ and $8x$ equivalent? ________.

Exercise 2 By choosing a value for x, show that the expressions $2+3x$ and $5x$ are not equivalent. (You can use $x=6$ if you like, or some other value.)

Answer:

Exercise 3 Explain the following ideas, and give an example for each.

a. equivalent expressions

b. like terms

c. numerical coefficient

Exercise 4

a. $-4y - 3y = (-4 - 3)y =$____________.

b. $-6st + 9st = ($__________$)st =$ ____________ .

Exercise 5 Delbert and Francine are collecting aluminum cans to recycle. They will be paid x dollars for every pound of cans they collect. At the end of three weeks, Delbert collected 23 pounds of aluminum cans, and Francine collected 47 pounds.

a. Write algebraic expressions for the amount of money Delbert made, and the amount Francine made.

Answers:

b. Write and simplify an expression for the total amount of money Delbert and Francine made from aluminum cans.

Answer:

Exercise 6 Combine like terms by adding or subtracting.

a. $-2x + 6x - x$

b. $3bc - (-4bc) - 8bc$

Answers: *a.*

b.

Exercise 7 Simplify $-5u - 6uv + 8uv + 9u$

Answer:

Exercise 8 One angle of a triangle is three times the smallest angle, and the third angle is 20° greater than the smallest angle. If the smallest angle is x, write an expression for the sum of the three angles in terms of x, and simplify.

Answer:

Exercise 9 Simplify $(32h - 26) + (-3 + 2h)$

Answer:

Exercise 10 Simplify $(3a - 2) - 2a - (5 - 2a)$.

Answer:

Exercise 11 The StageLights theater group plans to sell T-shirts to raise money. It will cost them $5x + 60$ dollars to print x T-shirts, and they will sell the T-shirts at $12 each. Write an expression for their profit from selling x T-shirts, and simplify.

Answer:

Exercise 12 The StageLights theater group from Exercise 11 would like to make $500 from the sale of T-shirts. How many T-shirts must they sell?

Answer:

Section 2.9 The Distributive Law

Exercise 1 Simplify each product.

a. $6(-2b)$ *b.* $-4(-7w)$

Answers *a.* *b.*

Exercise 2 Use the distributive law to simplify each expression.

a. $8(3y-6)$ *b.* $-3(7+5x)$

Answers: *a.* *b.*

Exercise 3 The length of a rectangle is 3 feet less than twice its width, w.

a. Write an expression for the length of the rectangle in terms of w.

b. Write and simplify an expression for the perimeter of the rectangle in terms of w.

Answers: *a.* *b.*

Exercise 4 Suppose the perimeter of the rectangle in Exercise 3 is 36 feet. Write and solve an equation to find the dimensions of the rectangle.

Answer:

Exercise 5 Solve the equation in Example 1 in the Reading assignment for this section. What is the speed of the current in the Columbia River?

Answer:

Section 2.10 Line of Best Fit

Exercise 1 Use Figure 2.55 to estimate values. Are you using interpolation or extrapolation?

a. How many millions of square kilometers of rain forest existed in 1960?
b. How many millions of square kilometers of rain forest existed in 1975?
c. How many millions of square kilometers of rain forest existed in 1995?
d. When were there 8.6 million square kilometers of rain forest?

Answers: *a.* *b.*

c. *d.*

Exercise 2 What does the regression line in Figure 2.58 predict for the height of a person whose (men's) shoe size is 9?

Answer:

Height ≈ ____________________ inches

Exercise 3 Is person A taller or shorter than predicted by the regression line in Figure 2.58? By how much?

Answer:

A is ____________________ than predicted by about _____________ inches.

☐ *Next, we'll use the height and shoe size data you collected in your class.*

1. Create a scatterplot for the data you collected. Use the grid in Figure 2.62. (Can you pick out which point corresponds to you?)

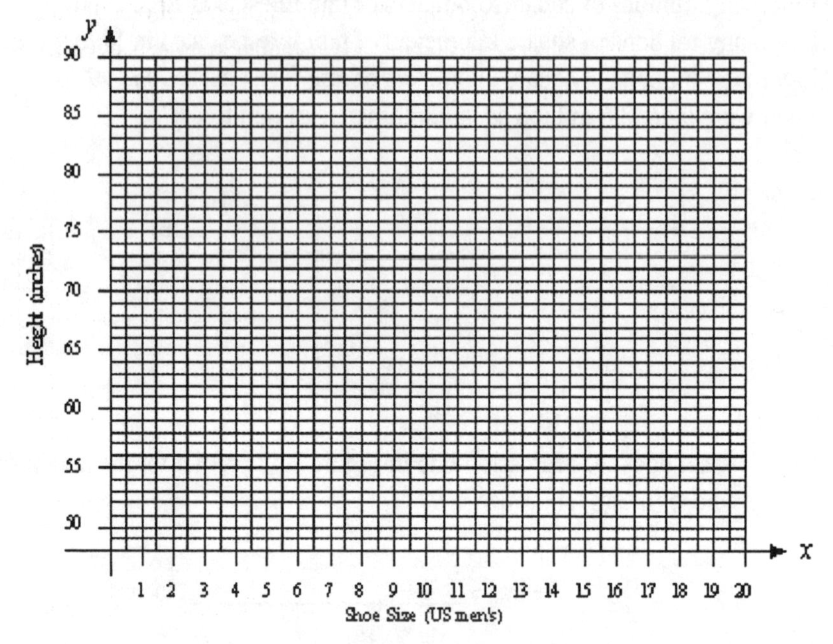

Figure 2.62

2. Draw a regression line for your scatterplot in Figure 2.62. According to your regression line, are you taller or shorter than predicted from your shoe size?

I am ____________________ than predicted by _______ inches.

3. According to your regression line, how tall do you expect a man to be if he wears size 18 shoes? Size 10 shoes?

Size 18 shoes: height ≈ _________ inches
Size 10 shoes: height ≈ _________ inches

Homework 2.10

☐ *Carefully draw a line that fits the given data, and use it to answer the question.*

3. A fruit punch recipe that makes 20 servings calls for 12 ounces of orange juice. How much orange juice is needed for 25 servings? (*Hint*: No orange juice is needed for 0 servings.) Use the grid below.

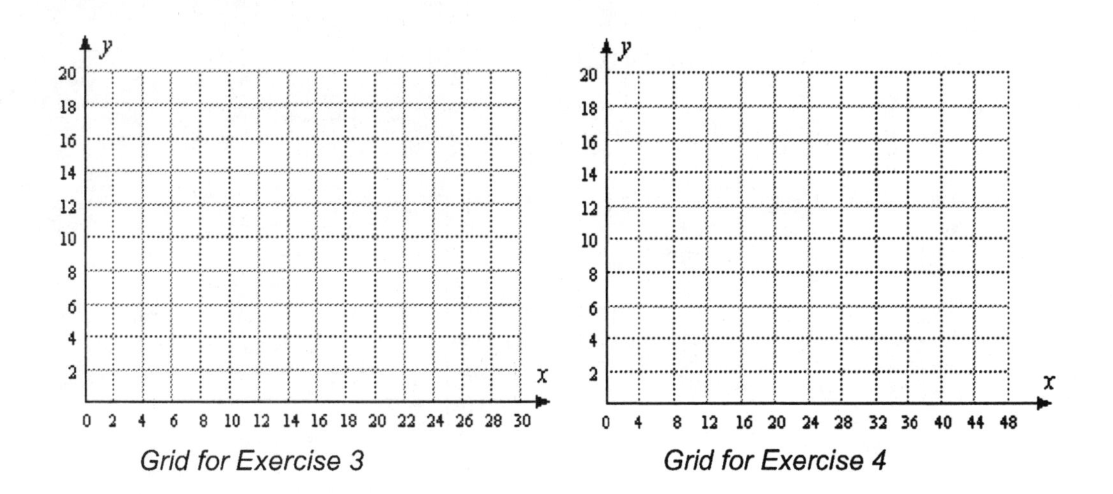

Grid for Exercise 3 *Grid for Exercise 4*

4. Delbert has a rose garden of area 24 square feet. He uses 9 small pails of compost for his garden. How much should he recommend for his neighbor's garden, which is 40 square feet? (*Hint*: No compost is needed for a garden of area 0 square feet.) Use the grid above.

5. Jennie keeps track of the number of gallons of gasoline used and the number of miles driven by her car between visits to the gas station. Use the grid below to plot the data.

Gallons	9.2	10.7	10.1	9.6	9.7	9.6	9.3
Miles	274.8	320.1	298.2	285.4	288.4	283.1	275.8

a. How far can Jennie expect to go on one tank of gas, if the tank holds 13.2 gallons?

b. How much gasoline will Jennie need to travel 310 miles?

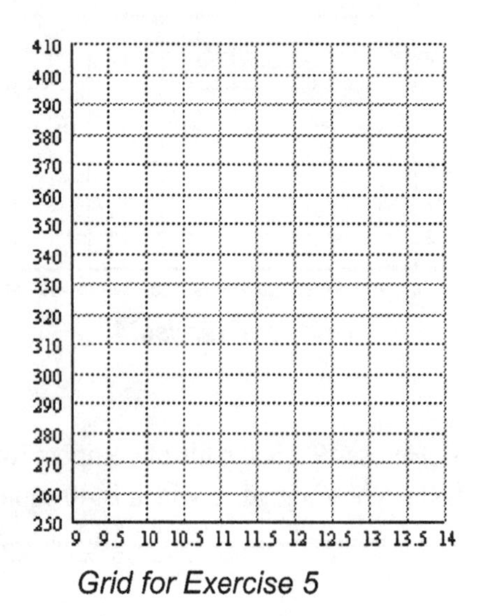

Grid for Exercise 5

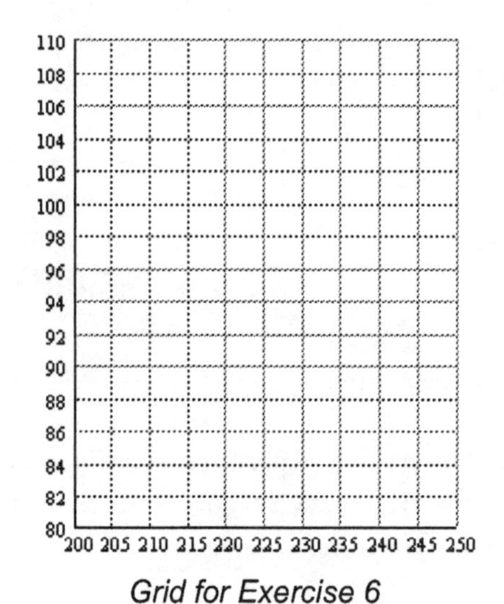

Grid for Exercise 6

6. Professor Martinez uses an overhead projector and an LCD panel to show her math class the display on her calculator. She measures the width of the image on the screen when the projector is at different distances from the screen. Use the grid above to plot the data.

Distance to screen (centimeters)	200	210	220	230	240
Width of image (centimeters)	87.6	91.8	96.4	100.6	104.8

a. Estimate the width of the image if the projector is 250 centimeters from the screen.

b. How far from the screen should Professor Martinez place the projector if she wants the image to be 95 centimeters wide?

7. Maita measured the length of a rubber band when it was supporting various objects of known weight. Use the grid below to plot the data.

Mass of object (ounces)	5	14.8	14	8
Length of rubber band (centimeter)	16.2	27.6	26.7	18.4

a. Estimate the length of the rubber band when it supports a 12 ounce weight.

b. Estimate the weight of an object that stretches the rubber band to a length of 20 centimeters.

c. What is the physical interpretation of the y-intercept of the regression line (if the graph is extended far enough to the left)?

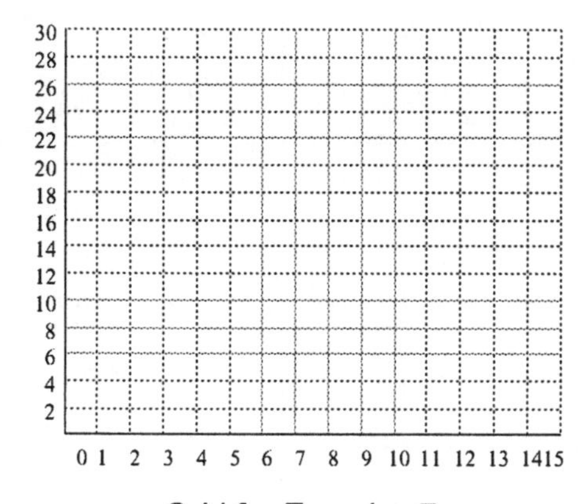

Grid for Exercise 7

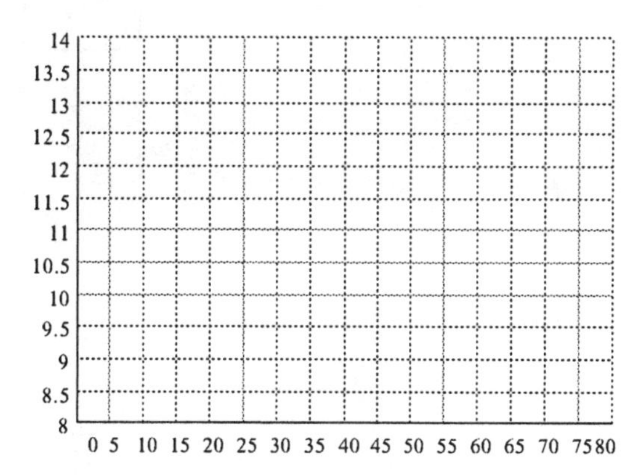

Grid for Exercise 8

8. Chia-ling measured the volume of a fixed quantity of gas at different temperatures. Use the grid above to plot the data.

Temperature (° Celsius)	20	40	60	80
Volume (liters)	10.3	11	11.7	12.4

a. Estimate the volume of the gas at a temperature of 37° Celsius.

b. Estimate the temperature needed to bring the gas to a volume of 11.5 liters.

c. What is the physical interpretation of the y-intercept of the regression line (if the graph is extended far enough to the right)?

d. What is the physical interpretation of the x-intercept of the regression line (if the graph is extended far enough downwards)?

11. In this exercise we will investigate how well a person's hand-span predicts his or her height. Measure the distance from the tip of your thumb to the end of your small finger when your fingers are fully extended. Record your hand-span and your height, in inches, in the table below. Obtain similar measurements for four more people.

Span (inches)					
Height (inches)					

a. Use the grid below to plot your data. Draw a regression line on the graph.

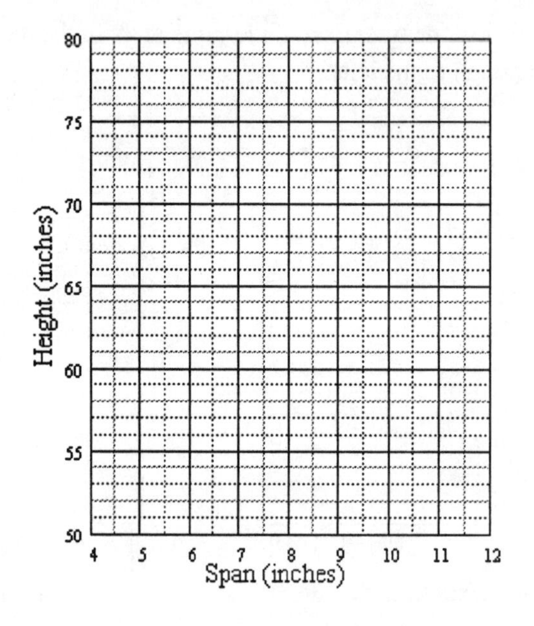

Grid for Exercise 11

b. Locate and label the point that corresponds to you. Are you taller or shorter than predicted by the regression line? By how much?

c. What does your regression line predict for the height of someone whose hand-span is 10.5 inches?

d. According to the regression line, what is the hand-span of someone who is 58 inches tall?

12. Does the number of course units you take determine the number of text books you will need? Poll eight students (including yourself) to see how many course units each is enrolled in this semester, and how many required text books each student has. Record the results in the table.

Number of Units								
No. of Required Books								

a. Use the grid below to plot your data. Draw a regression line on the graph.

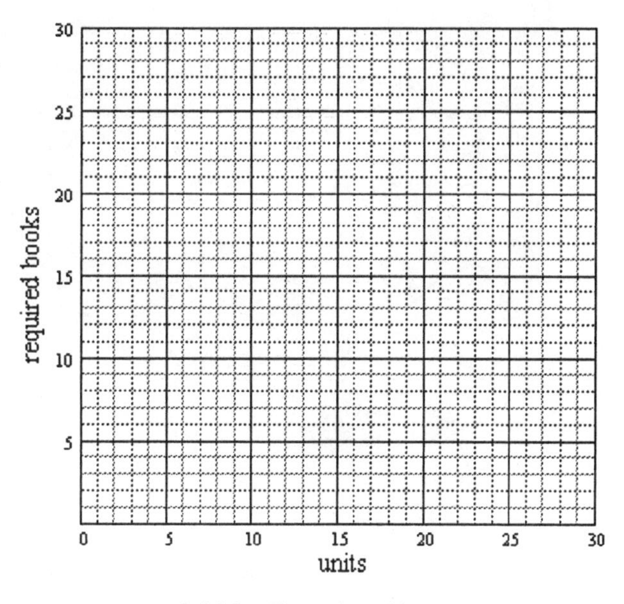

Grid for Exercise 12

b. Locate and label the point that corresponds to you. Do you have more or fewer required books than predicted by the regression line? By how many?

c. How many required books does your regression line predict for a student carrying 17 units?

d. According to the regression line, how many units is a student enrolled in if he or she has 6 required books?

Chapter 2 Summary and Review

41. The temperature in Maple Grove was $18°$F at noon, and it has been dropping ever since at a rate of $3°$F per hour.

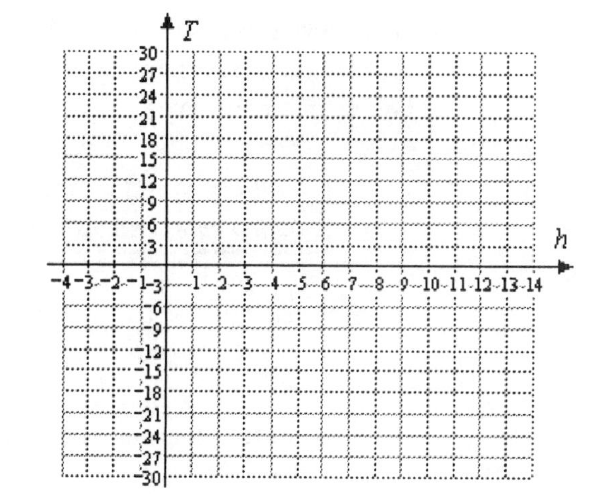

a. Fill in the table. T stands for the temperature h hours after noon. Negative values of h represent hours before noon.

h	-4	-2	0	1	3	5	8
T							

b. Write an equation for the temperature, T, after h hours.

c. Graph your equation.

d. What was the temperature at 10 am?

e. When will the temperature reach $-15°$F?

f. How much did the temperature drop between 3 pm and 9 pm?

☐ *Carefully draw a line that fits the given data, and use it to answer the questions.*

77. Gokhan was told that the rate at which crickets chirp depends on the air temperature. He records a cricket's chirp rate (in chirps per minute) at various temperatures. Use the grid below to plot the data.

Temperature (° Fahrenheit)	48.3	52.7	60.6	67.1	73.4	75.5
Chirp rate (chirps/minute)	32	52	84	108	132	142

a. Estimate how fast the cricket will chirp at a temperature of 70° F.

b. Estimate the temperature at which the cricket will chirp 100 times per minute.

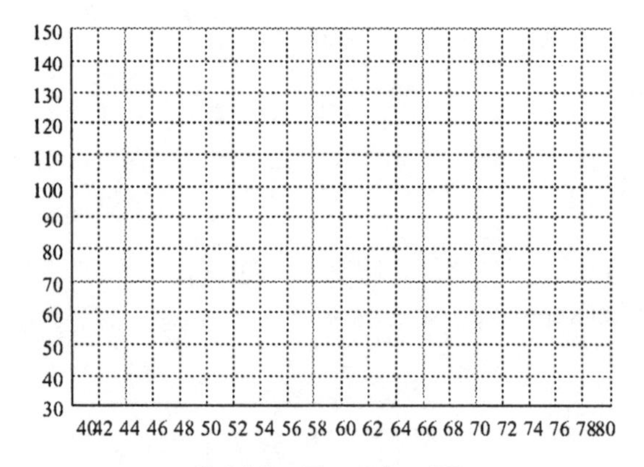

Grid for Exercise 77

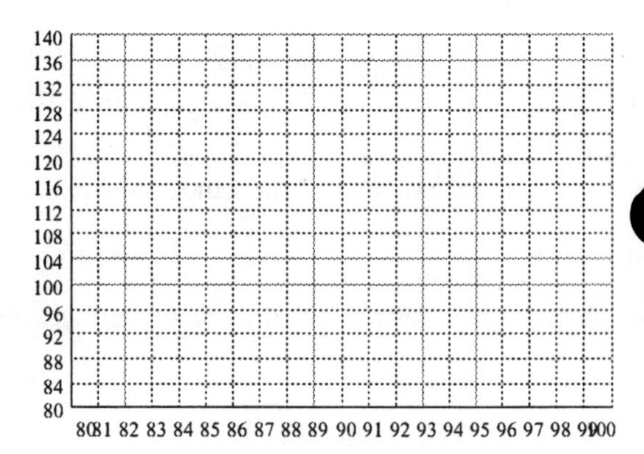

Grid for Exercise 78

78. The table below shows the total expenditures for all public colleges and universities in the United States for the years 1985 to 1993. The figures are given in billions of dollars. Use the grid above to plot the data.

Year	1985	1986	1987	1988	1989	1990	1991	1992	1993
Expenditure	81	86	90	92	96	102	106	108	111

a. Estimate when expenditures were at $74 billion.

b. Estimate when these data predict expenditures of $135 billion.

Chapter 3 Applications of Linear Equations

Section 3.1 Ratio and Proportion

Exercise 1 Find three different pairs of numbers whose ratio is $\frac{7}{4}$. (There are many answers.)

Answer:

Exercise 2 Bita made $240 for 25 hours of work last week. Express her pay as a ratio, and then as a rate.

Answer:

Exercise 3 Solve $\frac{q}{3.2} = \frac{1.25}{4}$.

Answer:

Number of Quarts	*Total Price (cents)*	$\frac{\textit{Total Price}}{\textit{Number of Quarts}}$
1	80	$\frac{80}{1} = 80$
2	160	$\frac{160}{2} =$
3	240	
4	320	

Table 3.1

Exercise 4 Show that the proportions

$$\frac{1000 \text{ cents}}{x \text{ quarts}} = \frac{80 \text{ cents}}{1 \text{ quart}} \quad \text{and} \quad \frac{x \text{ quarts}}{1000 \text{ cents}} = \frac{1 \text{ quart}}{80 \text{ cents}}$$

have the same solution, but the equation

$$\frac{1000 \text{ cents}}{x \text{ quarts}} = \frac{1 \text{ quart}}{80 \text{ cents}}$$

has a different solution.

Answer:

Exercise 5 Follow the steps suggested below to solve the problem: If Sarah can drive 390 miles on 15 gallons of gas, how much gas will she need to travel 800 miles?

Step 1 The variables *miles driven* and *gallons of gas used* are proportional. Make a ratio with these variables: $\frac{\text{miles}}{\text{gallons}}$

(We could also choose $\frac{\text{gallons}}{\text{miles}}$.)

Step 2 Create a proportion using the information in the problem: One ratio will use 390 miles and 15 gallons; the other will use the unknown quantity, x gallons, and 800 miles. Use the ratio from Step 1 to guide you.

Proportion:

Step 3 Solve the proportion.

Section 3.2 Similarity

Exercise 1 What fraction of a whole circle is the angle $7.2°$?

Answer:

Exercise 2 a. Solve the proportion to find the circumference of the earth in stades.

b. If one stade is about 157.5 meters, find the circumference of the earth in kilometers.

Answers a.

b.

Exercise 3 Which of the following figures are similar to the original?

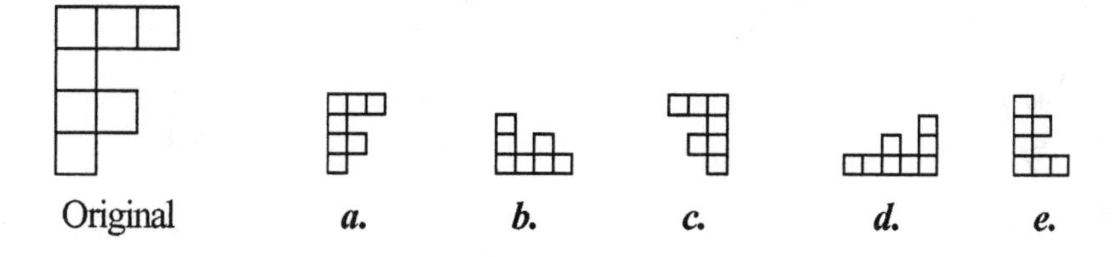

Answer:

Exercise 4 The following two figures are similar. Find the missing measurements, assuming that c is longer than 6.

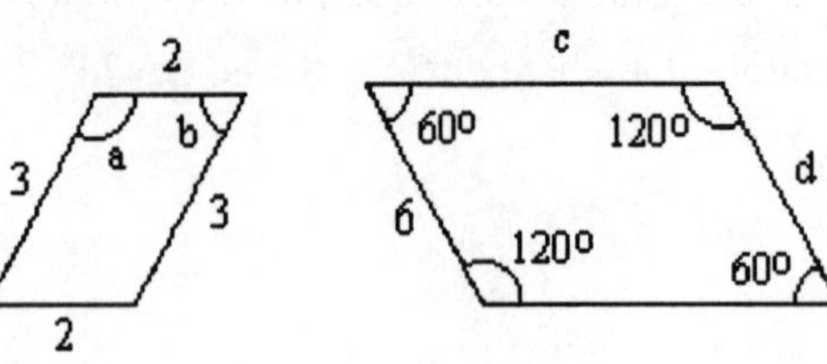

Answers: ***a.*** ***b.***

c. ***d.***

Exercise 5 Which of the following pairs of triangles are similar? Explain why or why not in each case.

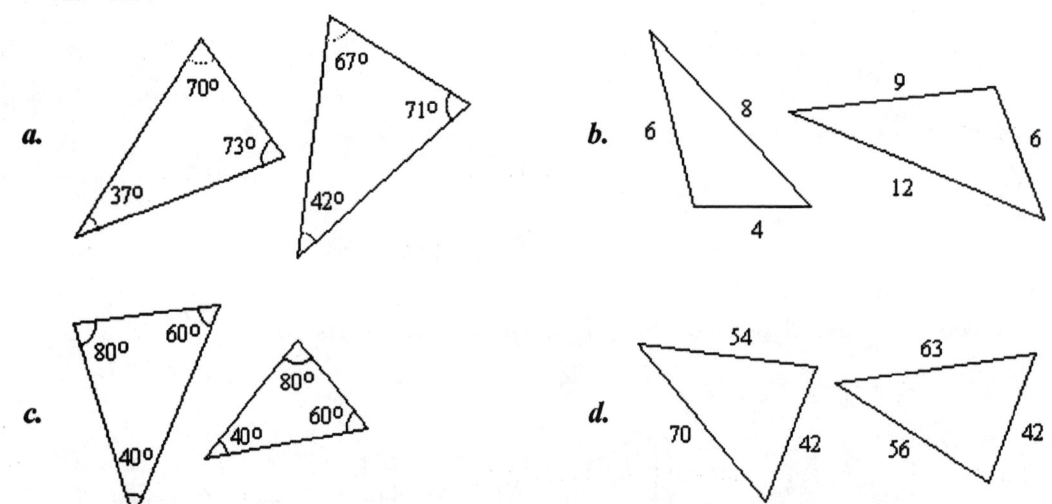

Answers: ***a.*** ***b.***

c. ***d.***

Exercise 6 Explain why the statement in the paragraph above is true: "If two pairs of corresponding angles in the triangles are equal, then the third pair must be equal also.

Answer:

Exercise 7 Find the value of h in Figure 3.7.

Answer:

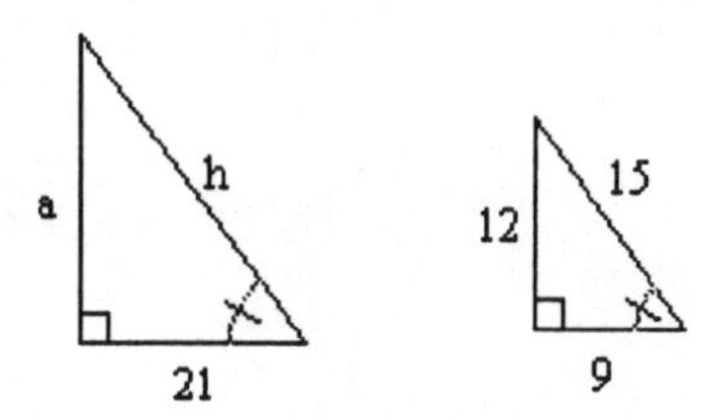

Figure 3.7

Exercise 8 Find the height of the building in Figure 3.8.

Answer:

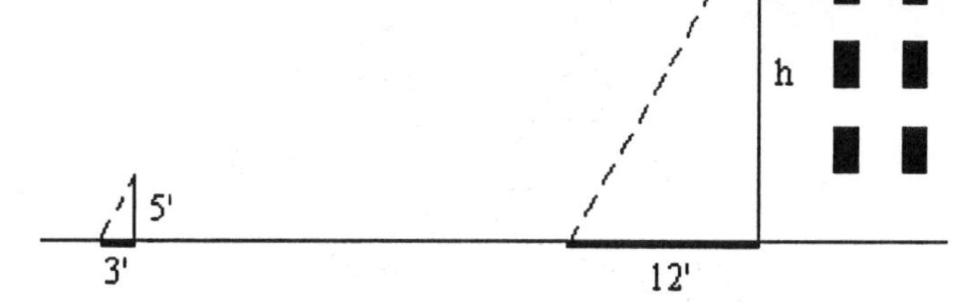

Figure 3.8

Exercise 9 Heather wants to know the height of a street lamp. She discovers one night that when she is 12 feet from the lamp, her shadow is 6 feet long. (See Figure 3.11.) Find the height of the street lamp.

Answer:

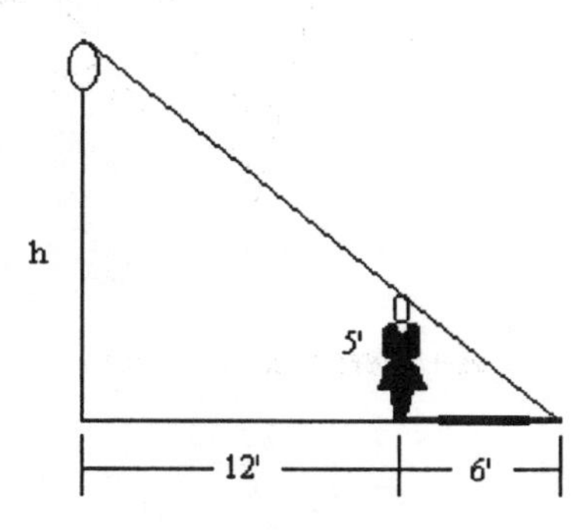

Figure 3.11

Section 3.3 Direct Variation

☐ a. *Complete the table to help you decide whether each pair of variables is proportional.*

b. *Graph the points on the grids provided.*

1. The table below shows the price, p, for g gallons of gasoline at the pump. Plot the data on the grid in Figure 3.17.

Gallons	*Total Price*	$\frac{\textit{Price}}{\textit{Gallons}}$
4	\$6.00	$\frac{6.00}{4} = ?$
6	\$9.00	$\frac{9.00}{6} = ?$
9	\$13.50	?
12	\$18.00	?
15	\$22.50	?

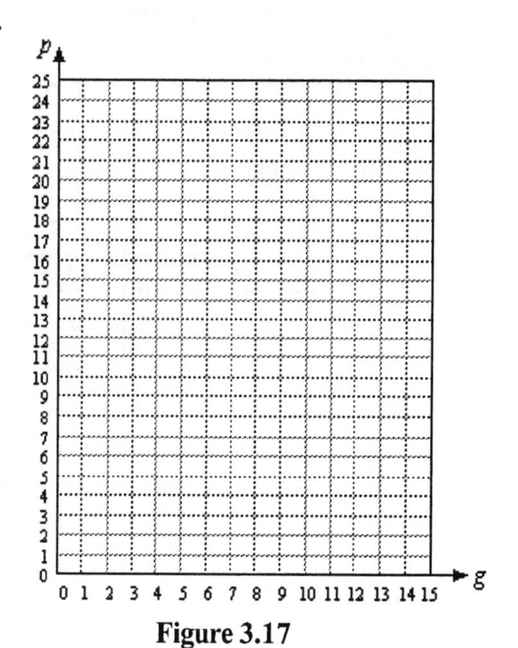

Figure 3.17

2. The table shows the growth in population, P, of a new suburb t years after it was built. Plot the data on the grid in Figure 3.18.

Years	*Population*	$\frac{\textit{People}}{\textit{Year}}$
1	10	$\frac{10}{1} = ?$
2	20	$\frac{20}{2} = ?$
3	40	?
4	80	?
5	160	?

P
200 190 180 170 160 150 140 130 120 110 100 90 80 70 60 50 40 30 20 10 0
0 1 2 3 4 5 6
t

Figure 3.18

3. At this point, can you make a conjecture (guess) about the graphs of proportional variables? To help you decide if your conjecture is true, continue with the graphs in #4 and #5.

Conjecture:

4. Tuition at Woodrow University is \$400 plus \$30 per unit.
   a. Write an equation for tuition, T, in terms of the number of units, u.
      $T =$
   b. Use your equation to fill in the second column of the table.
   c. Graph the equation on the grid in Figure 3.19.
   d. Are the variables proportional? Compute their ratios to decide.

Units	*Tuition*	$\frac{\textit{Tuition}}{\textit{Unit}}$
3		
5		
8		
10		
12		

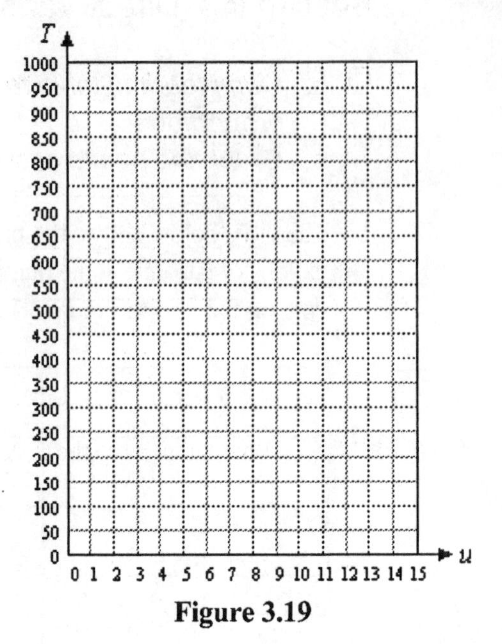

Figure 3.19

5. Anouk is traveling by train across Alaska at 60 miles per hour.
   a. Write an equation for the distance, D, Anouk has traveled in terms of hours, h.
      $D =$
   b. Use your equation to fill in the table.
   c. Graph the equation on the grid in Figure 3.20.
   d. Are the variables proportional? Compute their ratios to decide.

Hours	*Distance*	$\frac{\textit{Distance}}{\textit{Hour}}$
3		
5		
8		
10		
16		

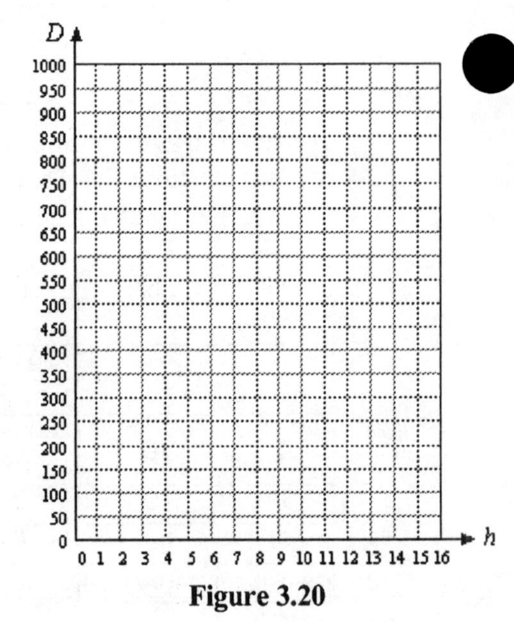

Figure 3.20

6. Can you revise your conjecture about the graphs of proportional variables so that it applies to the graphs on this page as well? *Hint*: Look at the two graphs of proportional variables. What is the y-intercept of both graphs?

Revised Conjecture:

Homework 3.3

☐ *Use the equations you wrote in Problems 9-12 to answer the questions.*

13. a. Fill in the table using your equation from Problem 9.

Time (hours)	*Distance (miles)*	*Time (hours)*	*Distance (miles)*
2		4	
3		6	
5		10	

b. What happens to the distance Everett travels when he doubles the time he bicycles?

14. a. Fill in the table using your equation from Problem 10.

Hours worked	*Paycheck*	*Hours worked*	*Paycheck*
4		12	
6		18	
8		24	

b. What happens to Josh's paycheck when he triples the number of hours he works?

15. a. Fill in the table using your equation from Problem 11.

Price	*Sales Tax*	*Price*	*Sales Tax*
12		6	
20		10	
30		15	

b. What happens to the sales tax when the price is cut in half?

16. a. Fill in the table using your equation from Problem 12.

Kola (ounces)	*Sugar (ounces)*	*Kola (ounces)*	*Sugar (ounces)*
20		5	
32		8	
64		16	

b. What happens to the amount of sugar when you take one-quarter of the amount of Kola?

☐ *Does your rule from Problem 17 hold if the variables are **not** proportional? Use problems 19 and 20 to help you decide.*

19. a. The cost of tuition at Walden College is given by the formula $T = 500 + 40u$, where u is the number of units you take. Is T proportional to u?

b. Fill in the table.

Units	*Tuition*	*Units*	*Tuition*
3		6	
5		10	
8		16	

c. Does doubling the number of units you take double your tuition?

20. a. The temperature in degrees Fahrenheit is given by $F = 1.8C + 32$, where C is the temperature in degrees Celsius. Is F proportional to C?

b. Fill in the table.

Celsius	*Fahrenheit*	*Celsius*	*Fahrenheit*
10		5	
30		15	
40		20	

c. If the Celsius temperature reading is reduced by half, is the Fahrenheit reading also reduced by half?

☐ *Problems 33-36 give the equations for three examples of direct variation. Complete the table of values for each equation, and graph them on the same coordinate axes.*

33. a. $y=2x$
    b. $z=3x$
    c. $w=1.5x$

x	y	z	w
-4			
-2			
0			
2			
4			

34. a. $y=-2x$
    b. $z=-2.5x$
    c. $w=-x$

x	y	z	w
-4			
-2			
0			
2			
4			

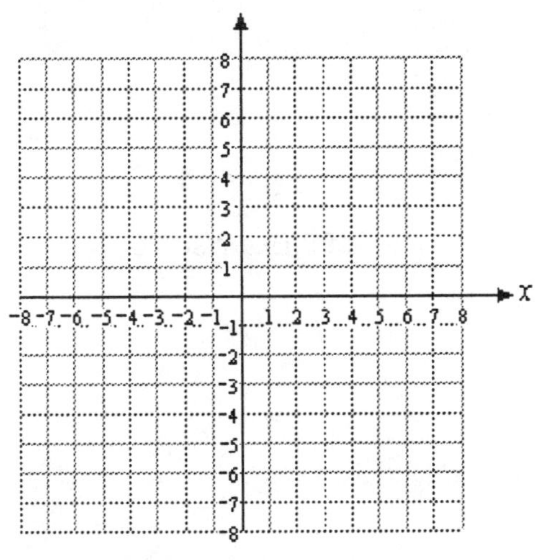

Grid for Exercise 33

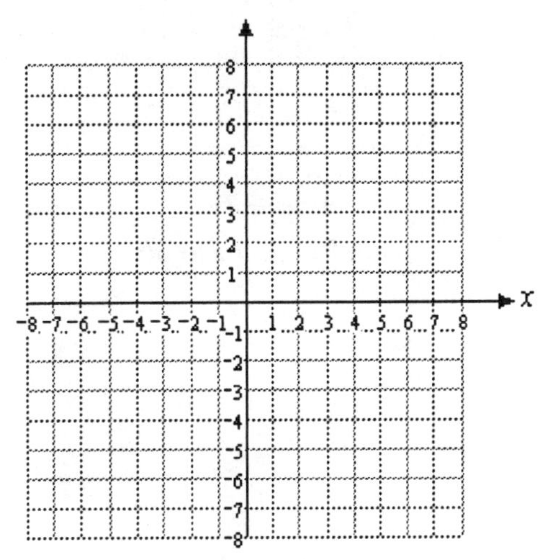

Grid for Exercise 34

35.

a. $y = -\frac{1}{3}x$

b. $z = -\frac{2}{3}x$

c. $w = -\frac{4}{3}x$

x	y	z	w
−6			
−3			
0			
3			
6			

36.

a. $y = \frac{1}{4}x$

b. $z = \frac{3}{4}x$

c. $w = \frac{7}{4}x$

x	y	z	w
−8			
−4			
0			
4			
8			

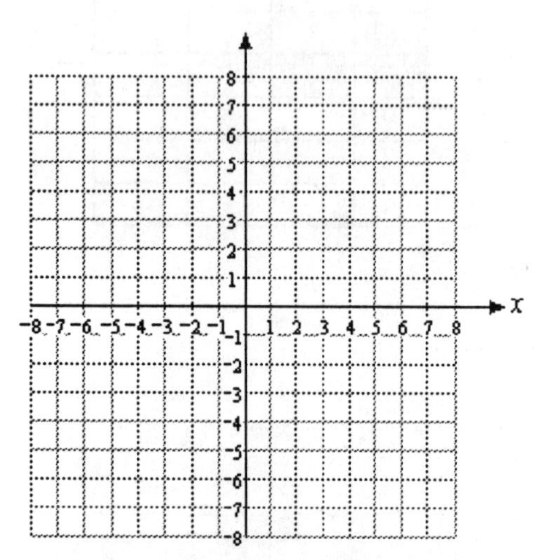

Grid for Exercise 35

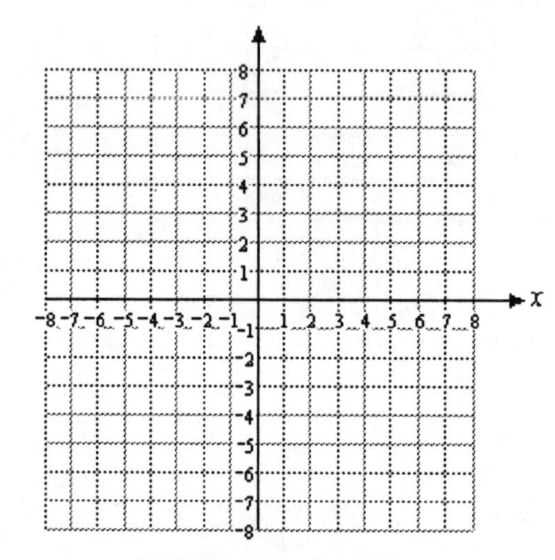

Grid for Exercise 36

Section 3.4 Slope

Exercise 1 Calculate the slope of the line in Figure 3.21 from the point $(10,74)$ to the point $(30,82)$. Illustrate Δy and Δx on the graph.

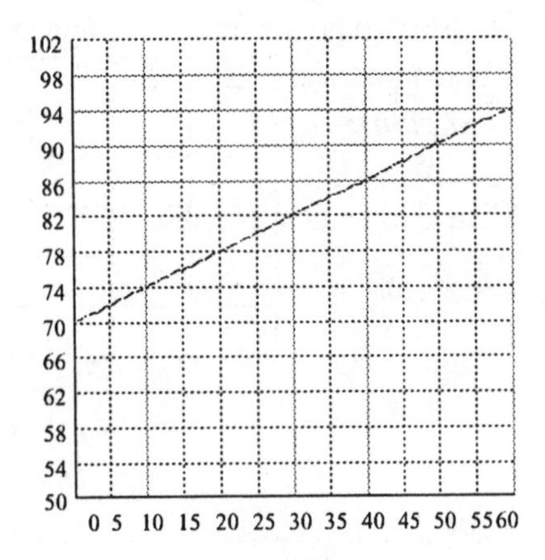

Figure 3.21

Answer:

Activity

In the following activity, we will calculate slopes for the graphs in Section 3.3. You should work in groups of three or four so that you can compare your results. Recall that to calculate the slope we choose two points on the graph and compute the ratio

$$\frac{\text{change in vertical coordinate}}{\text{change in horizontal coordinate}}.$$

Make sure to include the units with your ratios! Record your work below.

1. Choose two points from the graph in Figure 3.17, which shows the price, p, of gasoline in terms of the number of gallons, g, you buy.

First point: ***Second point:***

Change in vertical coordinates $\Delta p =$

Change in horizontal coordinates $\Delta g =$

Slope $\dfrac{\Delta p}{\Delta g} =$

2. Choose two points from the graph in Figure 3.18, which shows the population, P,of a new suburb t years after it was built.

First point: ***Second point:***

Change in vertical coordinates $\Delta P =$

Change in horizontal coordinates $\Delta t =$

Slope $\dfrac{\Delta P}{\Delta t} =$

3. Choose two points from the graph in Figure 3.19, which shows the cost, T,of tuition in terms of the number of units, u,taken.

First point: ***Second point:***

Change in vertical coordinates $\Delta T =$

Change in horizontal coordinates $\Delta u =$

Slope $\dfrac{\Delta T}{\Delta u} =$

4. Choose two points from the graph in Figure 3.20, which shows the distance, d,Anouk has traveled in terms of hours, h,elapsed.

First point: ***Second point:***

Change in vertical coordinates $\Delta d =$

Change in horizontal coordinates $\Delta h =$

Slope $\dfrac{\Delta d}{\Delta h} =$

5. **a.** Did everyone in your group get the same value for the slope of the graph in Figure 3.17?
b. Do you think you will always get the same value for the slope of this graph, no matter which two points you choose?

6. **a.** What about the graphs in Figures 3.18 through 3.20? Which of these graphs will give different values for the slope, depending on which points you choose?

b. What is different about this graph, compared to the other three graphs?

7. Remember that the slope is a *rate* of change. For each of your four graphs, answer the questions below. (The first one is done for you.)
a. What are the units of the rate?
b. What does the slope tell you about the variables involved?

Answers:

Figure 3.17: **a.** The units of $\frac{\Delta p}{\Delta g}$ are dollars per gallon.

b. The slope gives the cost of the gasoline in dollars per gallon. It is the rate of change of total price when the amount purchased increases.

Figure 3.18: **a.**

b.

Figure 3.19: **a.**

b.

Figure 3.20: **a.**

b.

Exercise 2 Find the slope of each line segment. (Each square counts for one unit.)

a.

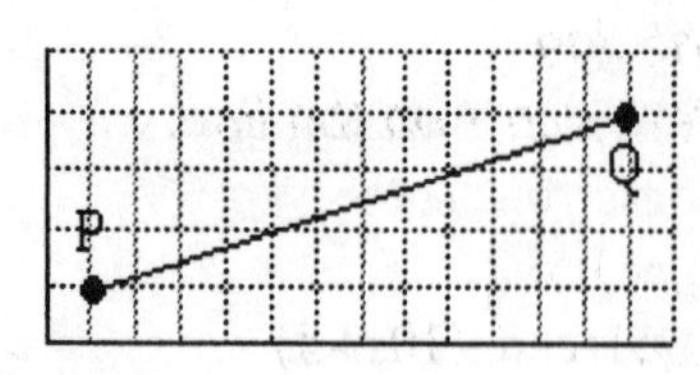

b.

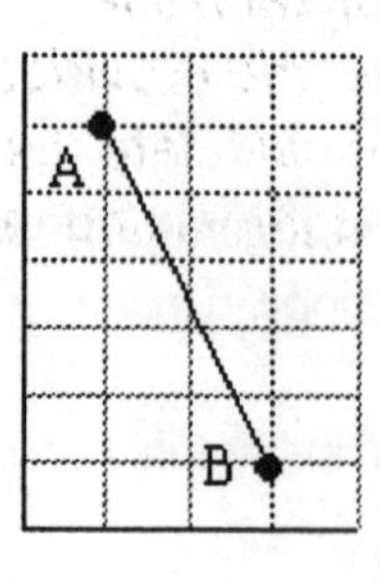

c.

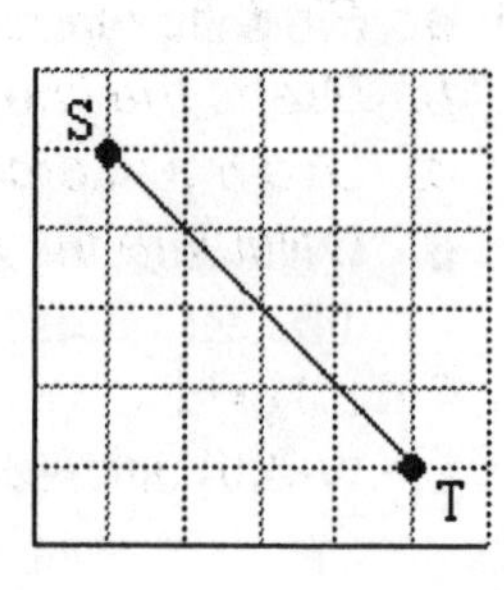

Answers: ***a.*** ***b.*** ***c.***

Exercise 3 In Exercise 2a, does it matter whether you move from P to Q or from Q to P to compute the slope of the line? Verify that you get the same answer if you move in the opposite direction.

Exercise 4 a. Which of the two graphs in Figure 3.28 appears steeper?

b. Compute the slopes of the two graphs. Which has the greater slope?

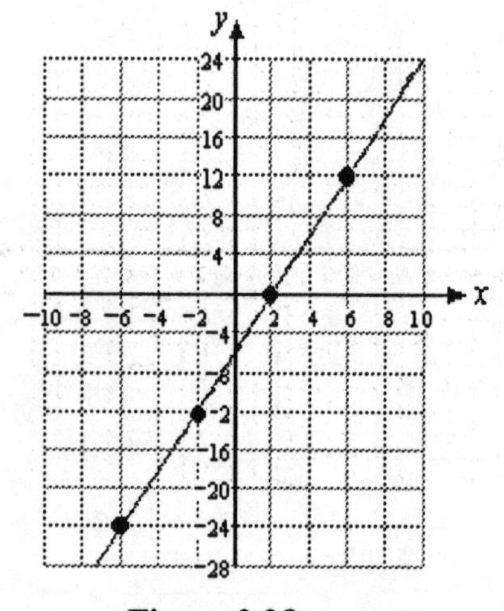

Figure 3.28a

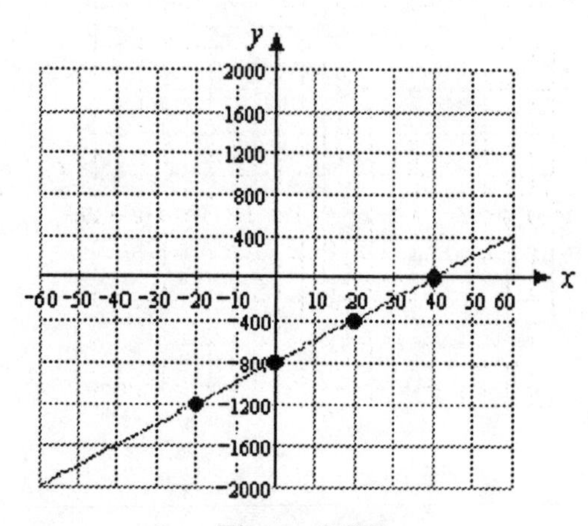

Figure 3.28b

Homework 3.4

☐ a. *Find the intercepts of each line.*

b. *Graph the line on the grid provided. Use the intercept method.*

c. *Use the intercepts to calculate the slope of the line.*

d. *Calculate the slope again using the suggested points on the line.*

9. $2x+3y=12$

   $(-3,6)$ and $(3,2)$

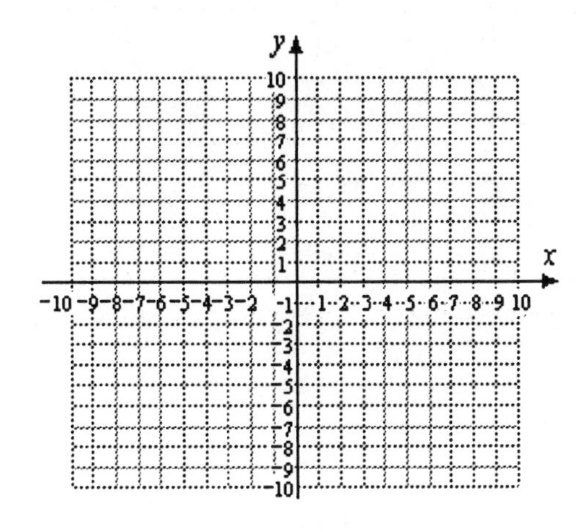

10. $3x+5y=15$

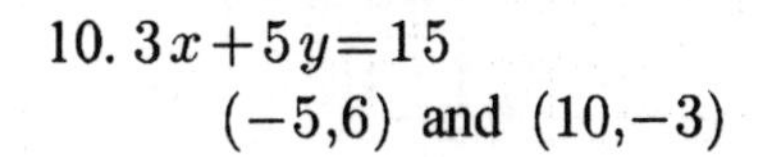

   $(-5,6)$ and $(10,-3)$

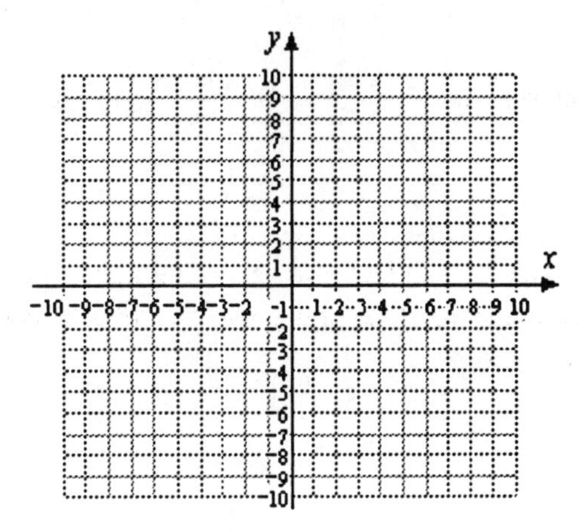

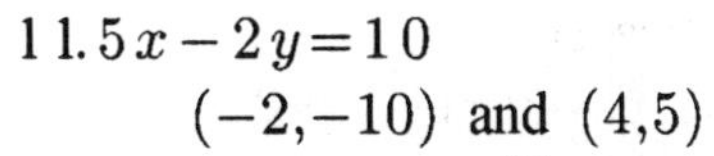

11. $5x-2y=10$

   $(-2,-10)$ and $(4,5)$

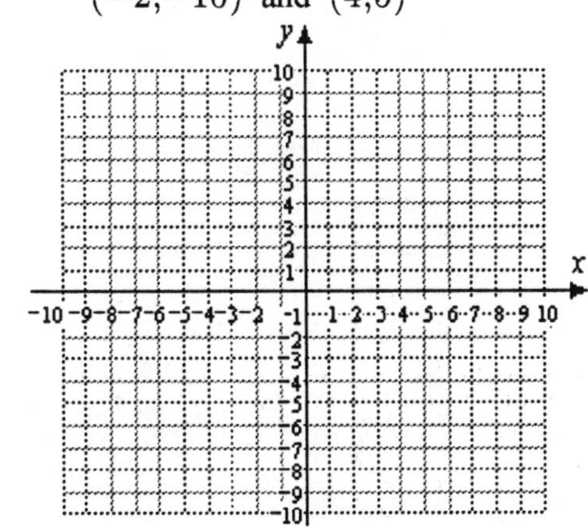

12. $4x-3y=12$

   $(-3,-8)$ and $(6,4)$

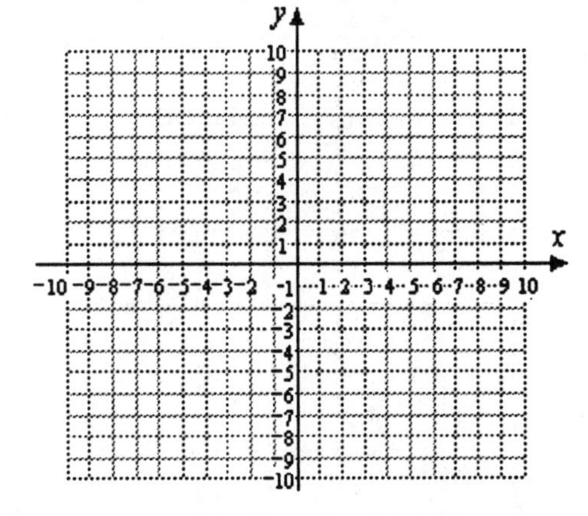

13. $x + y = 5$
$(-3,8)$ and $(8,-3)$

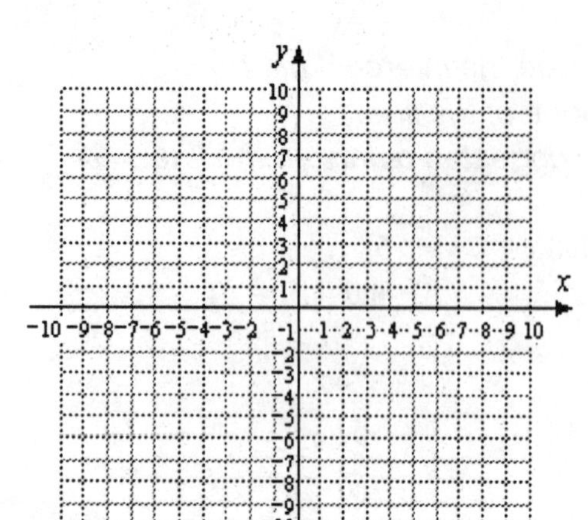

14. $x - y = 4$
$(6,2)$ and $(-2,-6)$

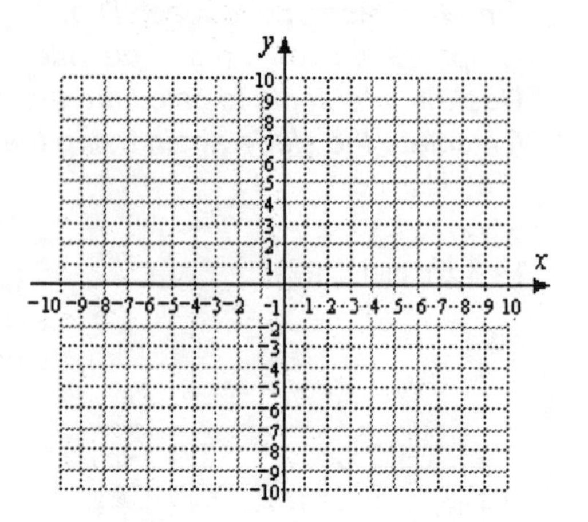

15. $x - 2y = 4$
$(6,1)$ and $(-4,-4)$

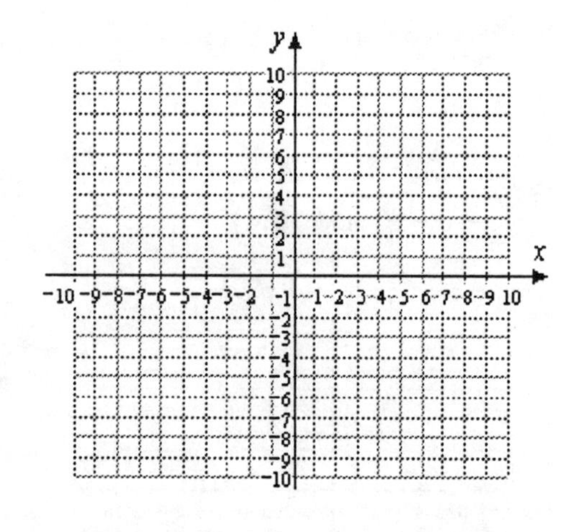

16. $3x + y = 6$
$(3,-3)$ and $(1,3)$

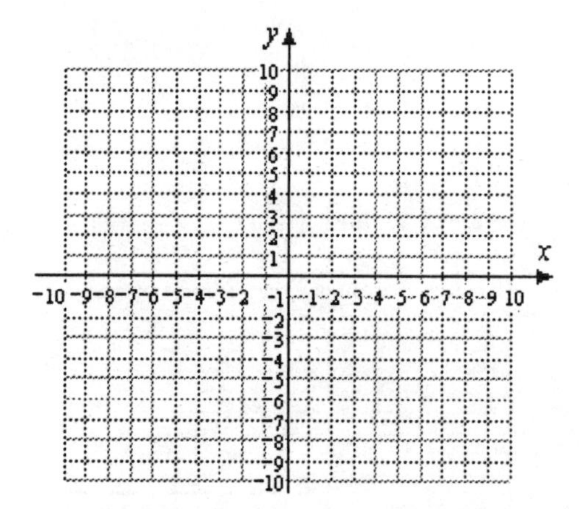

31. $y = -12x + 32$

x	-2	0	3	4
y				

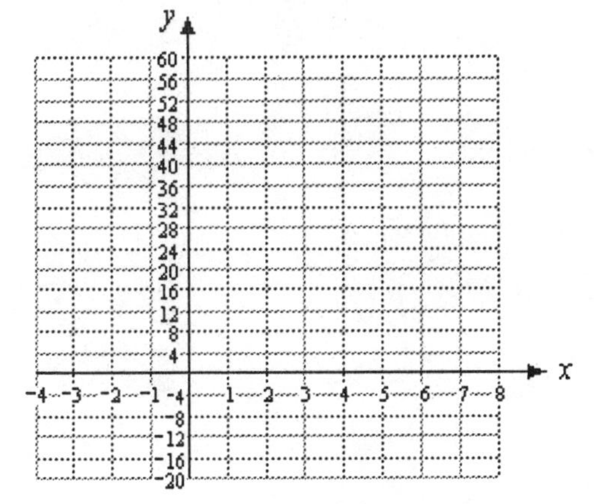

$m =$

32. $y = \frac{1}{2}x + 20$

x	-40	0	20	40
y				

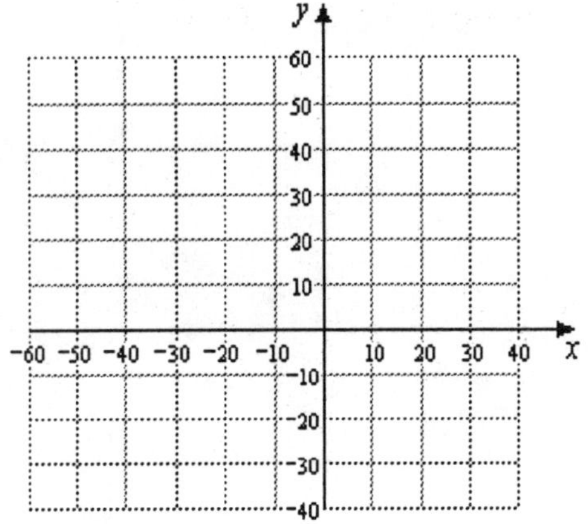

$m =$

Section 3.5 Slope-Intercept Form

Activity

We begin by considering how different slopes and different y-intercepts affect the graph of a line. In the following activity, we'll compare the graphs of several related equations.

1 a. Tuition at Woodrow University is \$400 plus \$30 per unit. Write an equation for tuition, W, in terms of the number of units, u.

$W =$

b. At Xavier College, the tuition, X, is \$200 plus \$30 per unit. Write an equation for X.

$X =$

c. At the Yardley Institute, the tuition, Y, is \$30 per unit. Write an equation for Y.

$Y =$

d. Fill in the table and graph all three equations on the grid in Figure 3.38.

u	W	X	Y
3			
5			
8			
10			
12			

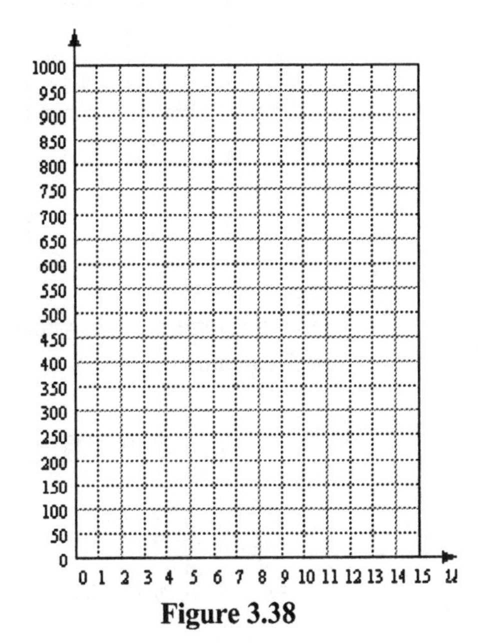

Figure 3.38

e. Find the slope and the y-intercept for each equation.

W: slope = y-intercept =

X: slope = y-intercept =

Y: slope = y-intercept =

f. How are your results from part (e) reflected in the graphs of the equations?

2 **a.** Anouk is traveling by train across Alaska at 60 miles per hour. Write an equation for the distance, A, Anouk has traveled in terms of hours, h.

$A =$

b. Boris is traveling by snowmobile at 30 miles per hour. Write an equation for Boris' distance, B.

$B =$

c. Chaka is traveling in a small plane at 100 miles per hour. Write an equation for Chaka's distance, C.

$C =$

d. fill in the table and graph all three equations on the grid in Figure 3.39.

h	A	B	C
3			
5			
8			
10			
16			

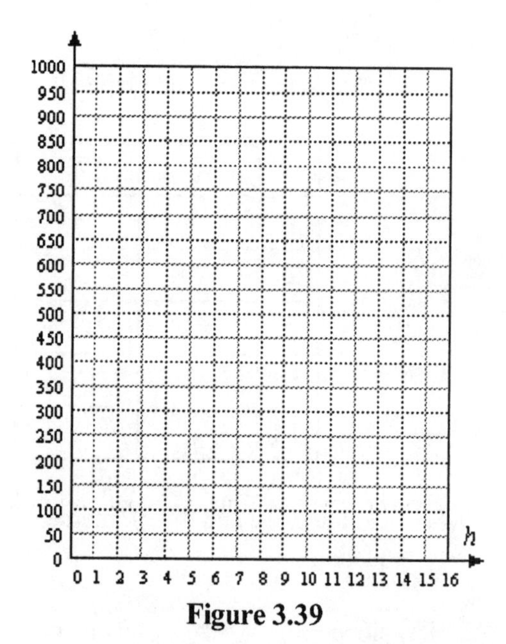

Figure 3.39

e. Find the slope and the y-intercept for each equation.

A: slope = y-intercept =

B: slope = y-intercept =

C: slope = y-intercept =

f. How are your results from part (e) reflected in the graphs of the equations?

Exercise 1 On Memorial Day week-end, Arturo drives from his home to a cabin on Diamond Lake. His distance from Diamond Lake after x hours of driving is given by the equation

$$y = 450 - 50x.$$

a. What are the slope and the y-intercept of the graph of this equation?
b. What do the slope and the y-intercept tell you about the problem?

Answers: ***a.***

b.

Exercise 2 Graph the equation $y = -3x + 4$ by the slope-intercept method. *Hint*: Write the slope as a fraction,

$$m = \frac{\Delta y}{\Delta x} = \frac{-3}{1}.$$

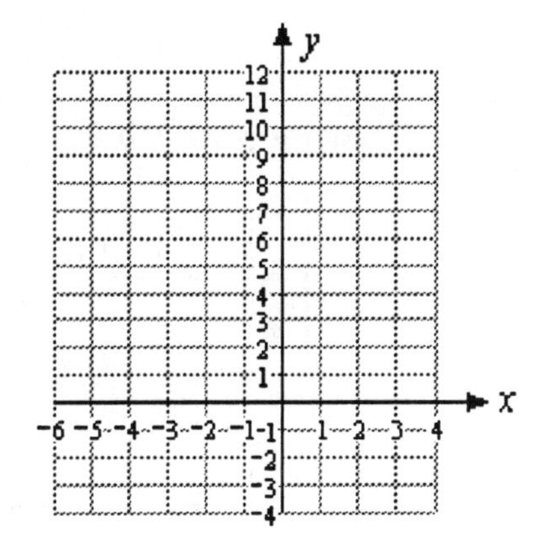

Exercise 3 Write an equation for the line whose y-intercept is $(0,4)$ and whose slope is -3.

Answer:

Homework 3.5

☐ *In Problems 1-8, compare the three equations.*

a. Fill in the y-values in the tables and graph the line.

b. Choose two points on each line and compute its slope.

1\. I. $y = 2x - 6$

x	−1	0	1	2	3
y					

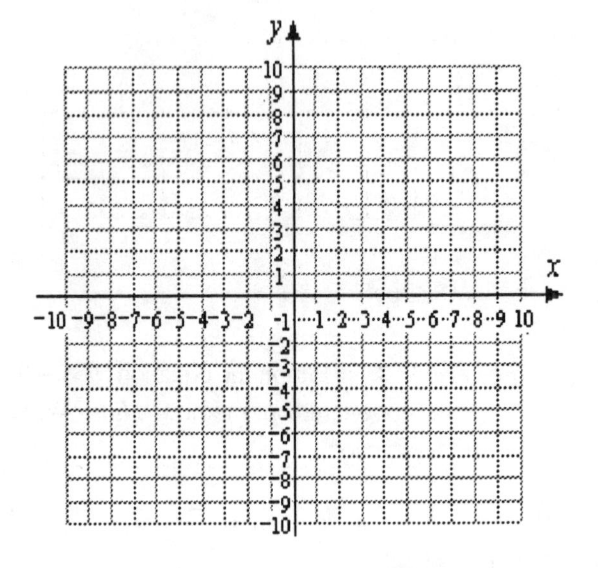

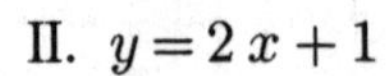

II. $y = 2x + 1$

x	−1	0	1	2	3
y					

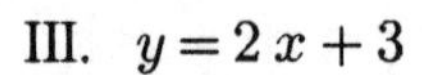

III. $y = 2x + 3$

x	−1	0	1	2	3
y					

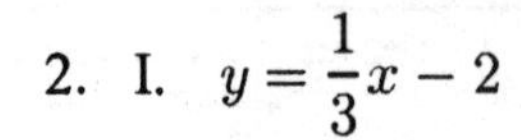

2\. I. $y = \frac{1}{3}x - 2$

x	−6	−3	0	3	6
y					

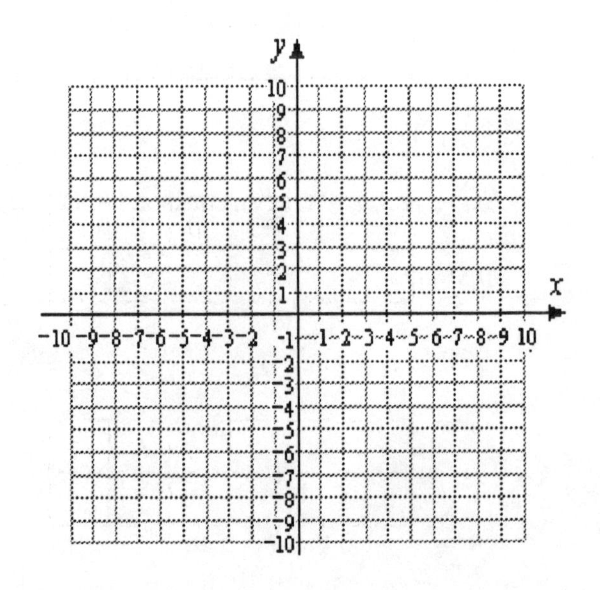

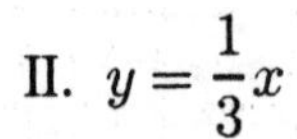

II. $y = \frac{1}{3}x$

x	−6	−3	0	3	6
y					

III. $y = \frac{1}{3}x + 4$

x	−6	−3	0	3	6
y					

3. I. $y = \frac{-3}{2}x - 4$

x	−6	−4	−2	0	2
y					

II. $y = \frac{-3}{2}x + 2$

x	−6	−4	−2	0	2
y					

III. $y = \frac{-3}{2}x + 6$

x	−6	−4	−2	0	2
y					

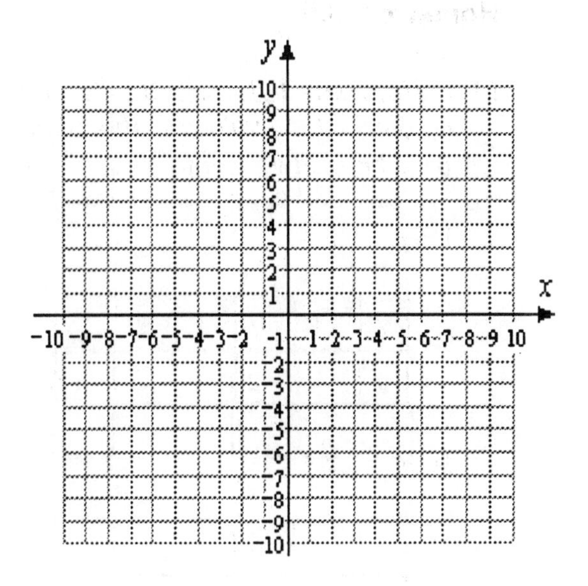

4. I. $y = -x - 1$

x	−4	−2	0	2	4
y					

II. $y = -x$

x	−4	−2	0	2	4
y					

III. $y = -x + 3$

x	−4	−2	0	2	4
y					

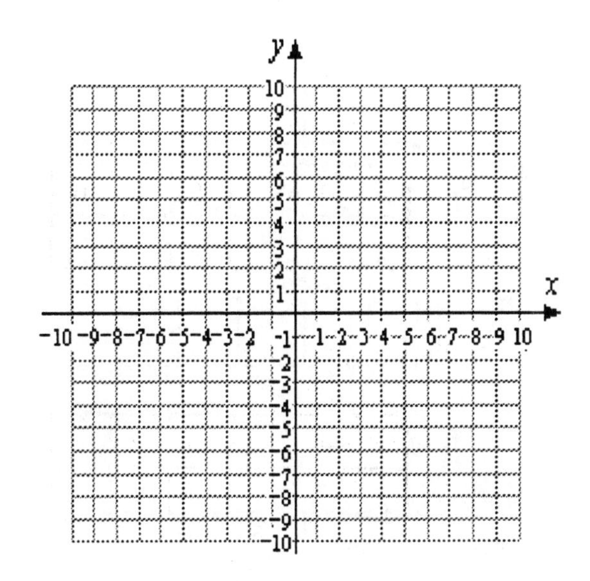

5. I. $y = \frac{1}{4}x + 2$

x	-4	-2	0	2	4
y					

II. $y = \frac{1}{2}x + 2$

x	-4	-2	0	2	4
y					

III. $y = x + 2$

x	-4	-2	0	2	4
y					

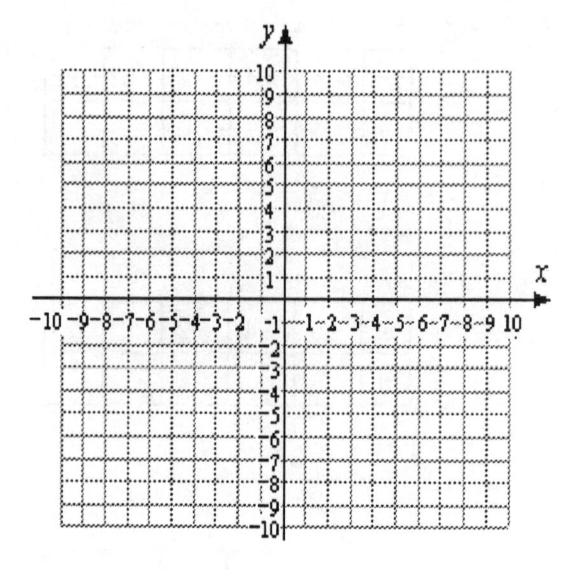

6. I $y = 2x + 2$

x	-4	-2	0	2	4
y					

II. $y = 4x + 2$

x	-4	-2	0	2	4
y					

III. $y = \frac{5}{2}x + 2$

x	-4	-2	0	2	4
y					

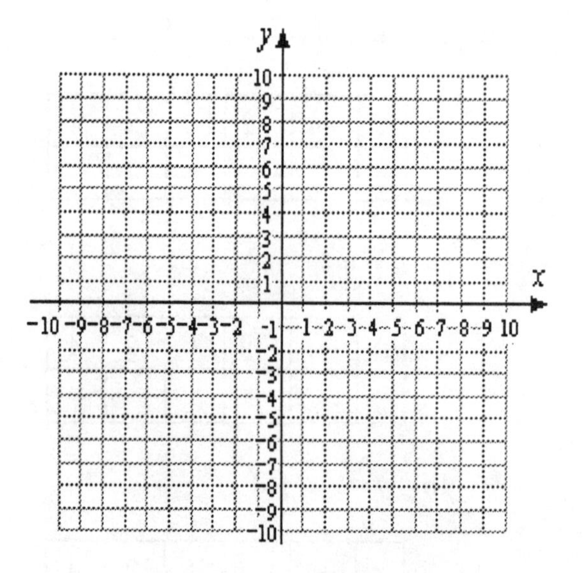

7. I. $y = -3x - 2$

x	-6	-3	0	3	6
y					

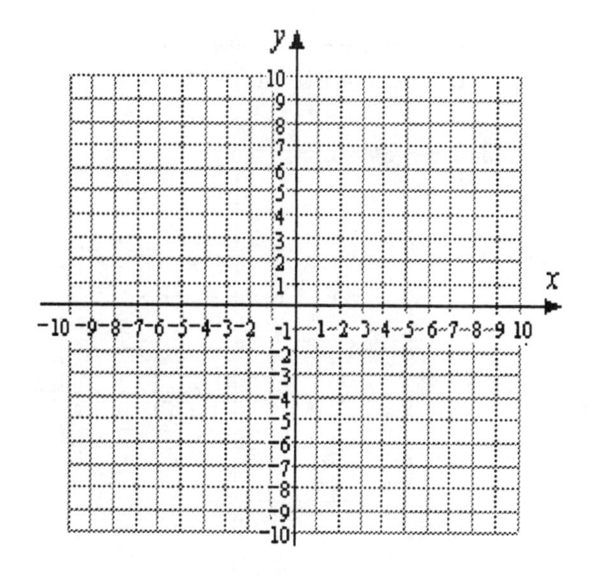

II. $y = -2x - 2$

x	-6	-3	0	3	6
y					

III $y = \dfrac{-5}{3}x - 2$

x	-6	-3	0	3	6
y					

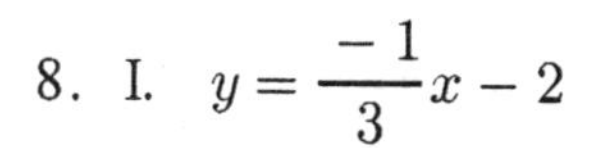

8. I. $y = \dfrac{-1}{3}x - 2$

x	-6	-3	0	3	6
y					

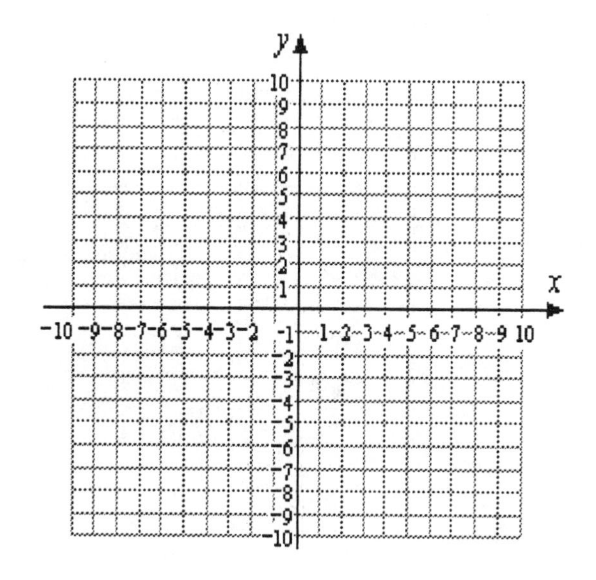

II. $y = \dfrac{-2}{3}x - 2$

x	-6	-3	0	3	6
y					

III $y = -x - 2$

x	-6	-3	0	3	6
y					

☐ *a. Find the intercepts of the graph,*
b. Graph the line,
c. Compute the slope of the line,
d. Put the equation in slope-intercept form.

23. $3x+4y=12$

a.

x	y
0	
	0

c. $\dfrac{\Delta y}{\Delta x} =$

d. $y =$

24. $2x-3y=6$

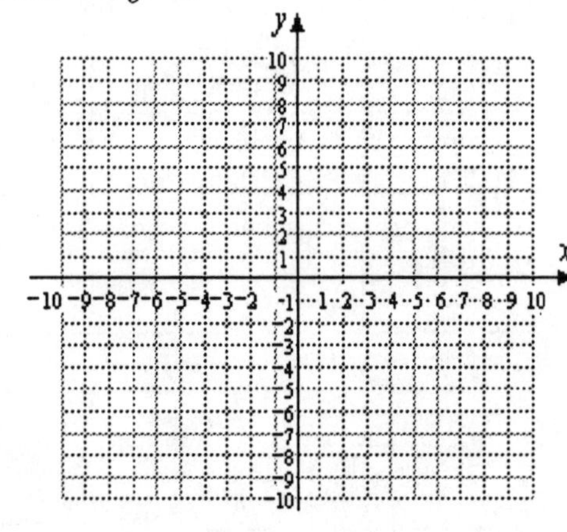

a.

x	y
0	
	0

c. $\dfrac{\Delta y}{\Delta x} =$

d. $y =$

25. $y + 3x - 8 = 0$

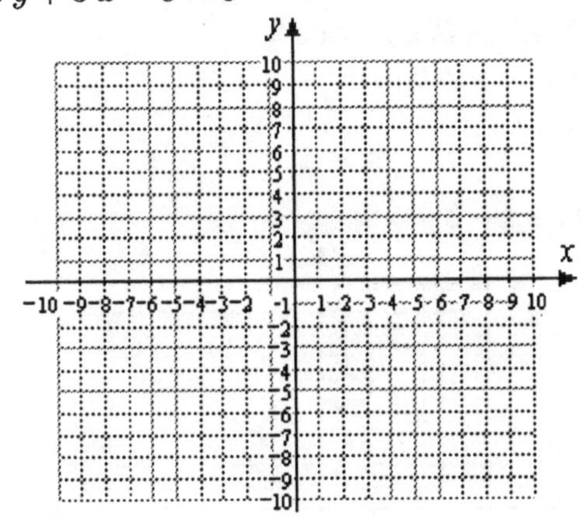

a.

x	y
0	
	0

c. $\dfrac{\Delta y}{\Delta x} =$

d. $y =$

26. $5 - x - 2y = 0$

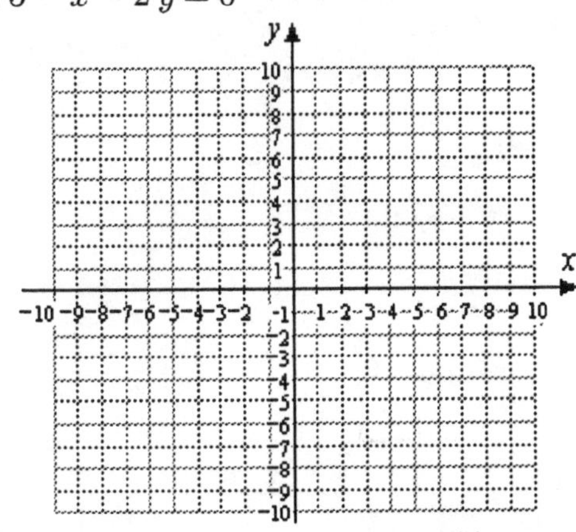

a.

x	y
0	
	0

c. $\dfrac{\Delta y}{\Delta x} =$

d. $y =$

☐ a. *In Exercises 27-30, put the equation in slope-intercept form.*
   b. *What is the y-intercept of each line? What is its slope?*
   b. *Use the slope to find four more points on the line.*
   c. *Graph the line.*

27. $3x - 5y = 0$

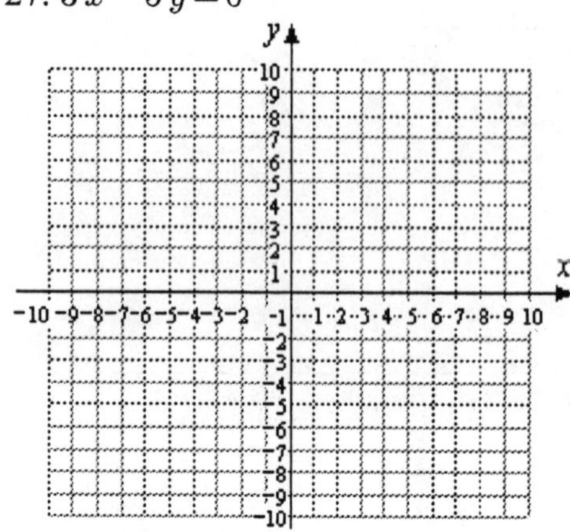

28. $2x + 3y = 0$

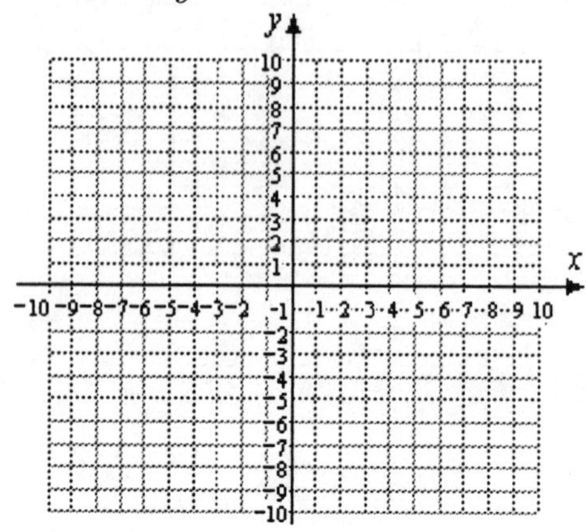

29. $5x + 4y = 0$

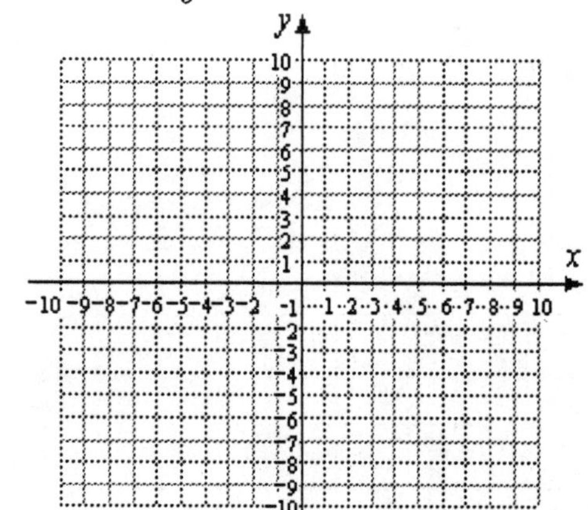

30. $5x - 2y = 0$

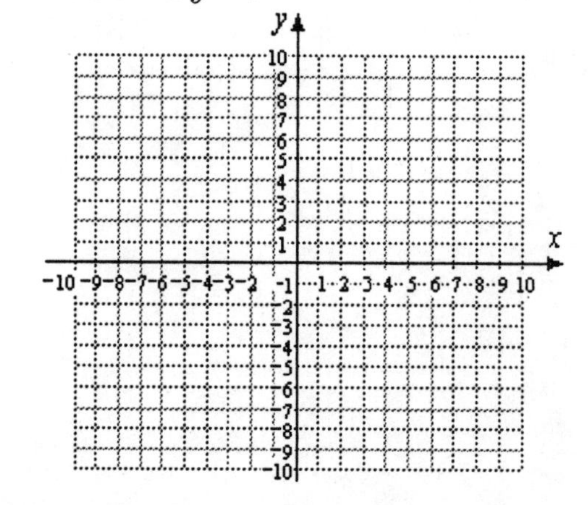

Midchapter Review

☐ a. *Find the x- and the y-intercepts of the line.*
   b. *Find the slope of the line.*
   c. *Write the equation of the line in slope-intercept form.*
   d. *Sketch the line.*

27. $5x - 4y = 20$

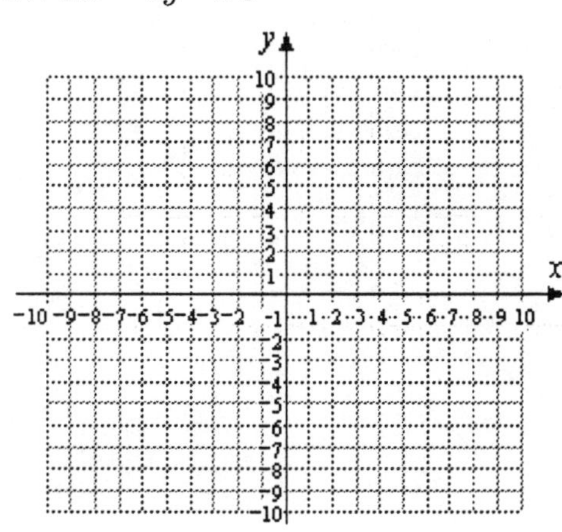

28. $\frac{y}{5} - \frac{x}{3} = 1$

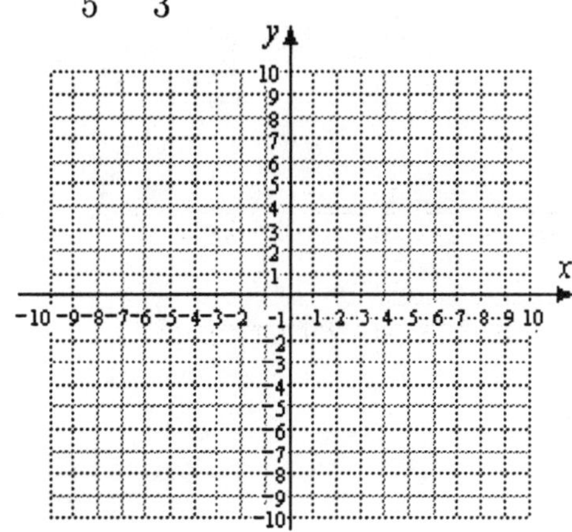

☐ *Graph the line with the given slope and passing through the given point.*

33. $m = -\frac{3}{4}$, $(2, -1)$

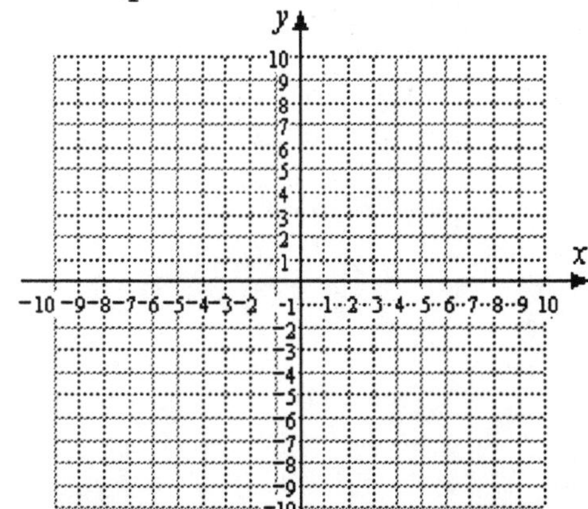

34. $m = \frac{1}{3}$, $(0, -3)$

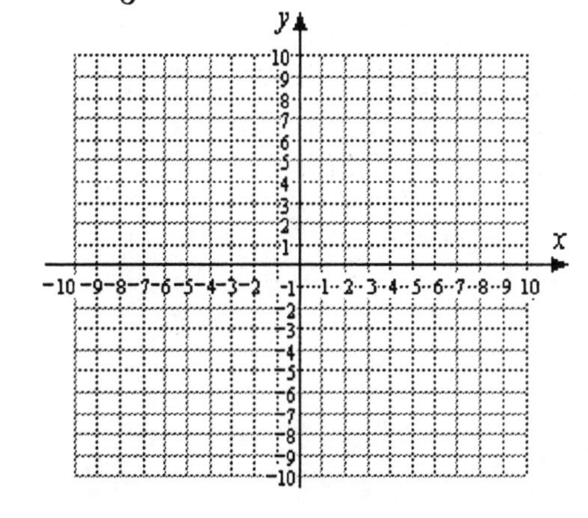

Section 3.6 Systems of Linear Equations

Exercise 1 Verify that the point $(50,1300)$ is a solution of each of the equations

$$S = 6t+1000,$$
$$E=2t+1200.$$

Answer:

Exercise 2 Decide whether the ordered pair $(2,3)$ is a solution to the system

$$3x-4y=-6$$
$$x+2y=-4.$$

Answer:

Exercise 3 Fill in the table and graph the equation for Ruth's distance, $d=65t$.

t	0	5	10
d			

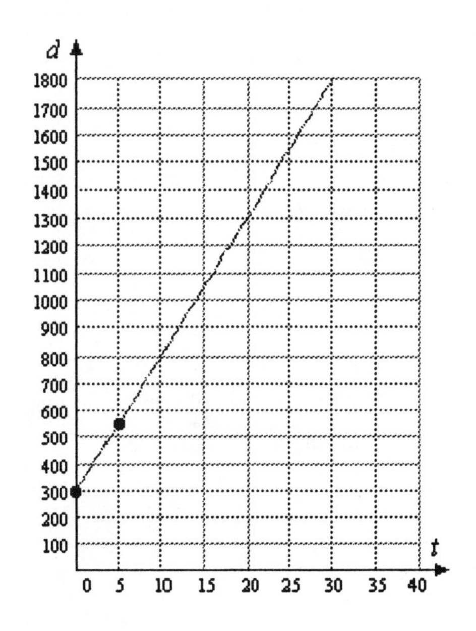

Exercise 4 Locate the point where the two graph intersect, and answer the questions:

a. What are the coordinates of the intersection point?

b. What does the t-coordinate of the intersection point tell you about the problem?

c. What does the d-coordinate of the point tell you?

d. Verify that the intersection point is a solution of both equations in the system.

Exercise 5 Solve the system

$$y = x - 3$$
$$y = 2x - 8.$$

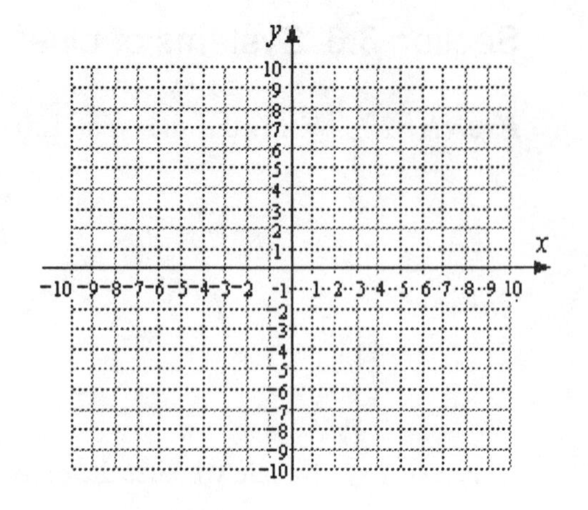

Use the slope-intercept method to graph each line. Verify that your solution satisfies both equations in the system.

$y = x - 3$ $\qquad\qquad$ $y = 2x - 8$

$b =$ $\qquad\qquad$ $b =$

$m = \dfrac{\Delta y}{\Delta x} =$ $\qquad\qquad$ $m = \dfrac{\Delta y}{\Delta x} =$

Exercise 6 Solve the system

$$3x = 2y + 6$$
$$y = \frac{3}{2}x - 1.$$

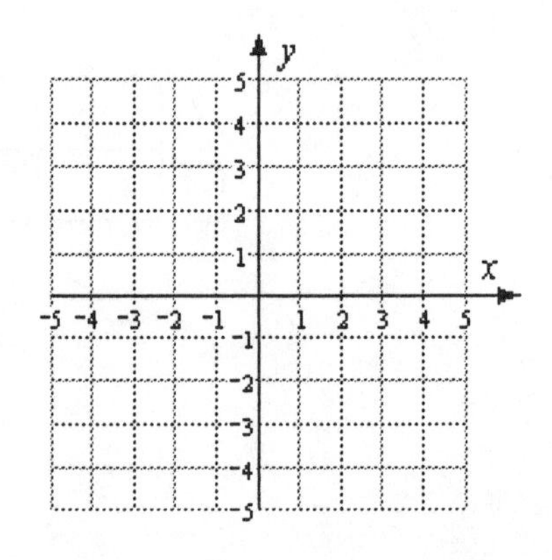

Use the intercept method to graph the first equation, and the slope-intercept method to graph the second equation.

Example 4 Allen has been asked to design a rectangular plexiglass plate whose perimeter is 28 inches and whose length is three times its width. What should the dimensions of the plate be?

Solution The "dimensions" of a rectangle are its length and width, so we'll let x represent the width of the plate and y represent its length. We must write two equations about the length and width of the plate. We know a formula for the perimeter of a rectangle, $P = 2l + 2w$, so our first equation is

$$2x + 2y = 28.$$

We also know that the length is three times the width, or

$$y = 3x.$$

These two equations make a system. We'll graph them both on the grid in Figure 3.47. Use the intercept method to graph the first equation, $2x + 2y = 28$:

x	y
0	
	0

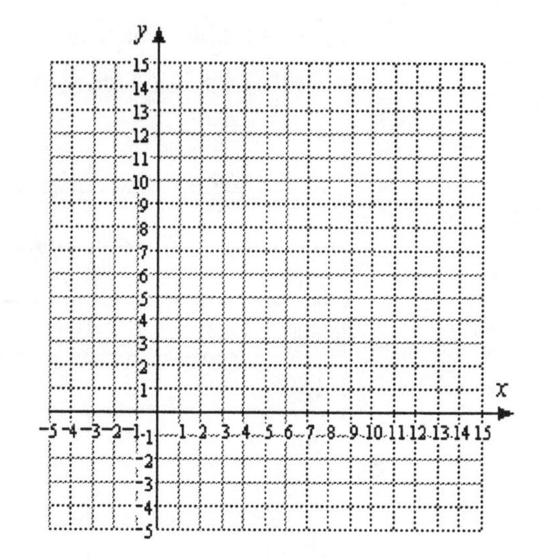

Figure 3.47

Use the slope-intercept method to graph $y = 3x$:

$b =$ $m = \dfrac{\Delta y}{\Delta x} =$

Homework 3.6

29. Delbert has accepted a sales job and is offered a choice of two salary plans. Under Plan A he receives $20,000 a year plus a 3% commission on his sales. Plan B offers a $15,000 annual salary plus a 5% commission.

   a. Let x stand for the amount of Delbert's sales in one year. Write equations for his total annual earnings under each plan.

   b. Fill in the table, where x is given in thousands of dollars.

   c. Graph both of your equations on the grid. (Both axes are scaled in thousands of dollars.)

x	*Earnings Under Plan A*	*Earnings Under Plan B*
0		
50		
100		
150		
200		
250		
300		
350		
400		

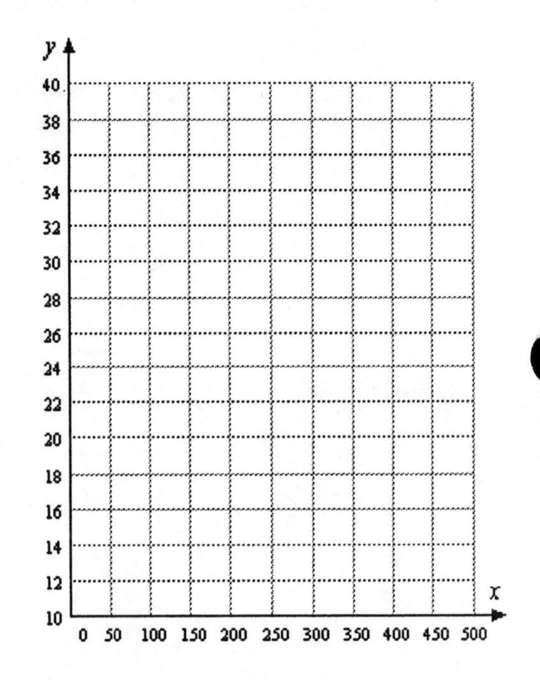

   d. For what sales amount do the two plans result in equal earnings for Delbert?

   e. Under what circumstances should Delbert prefer Plan A?

   f. If Delbert chooses plan B, how much must he sell in order to make more than $30,000 a year? What if he chooses plan A?

   g. Verify your answers to part (f) by writing and solving inequalities.

30. Francine wants to join a health club and has narrowed it down to two choices. The Sportshaus charges an initiation fee of $500 and $10 per month thereafter. Fitness First has an initiation fee of $50 and charges $25 per month.

   a. Let x stand for the number of months Francine uses the health club. Write equations for the total cost of each health club for x months.

   b. Fill in the table for the total cost of each club.

   c. Graph both equations on the grid.

x	*Sportshaus Total Cost*	*Fitness First Total Cost*
6		
12		
18		
24		
30		
36		
42		
48		

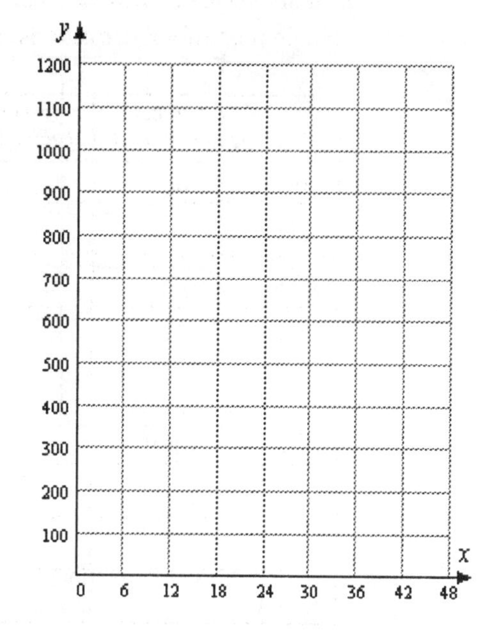

   d. When will the total cost of the two health clubs be equal?

   e. Under what circumstances should Francine prefer Fitness First?

   f. If Francine chooses Fitness First, when will the total cost exceed $650? What if she chooses the Sporthaus?

   g. Verify your answers to part (f) by writing and solving inequalities.

31. Orpheus Music plans to manufacture clarinets for schools. Their start-up costs are \$6000, and each clarinet costs \$60 to make. They plan to sell the clarinets for \$80 each.

a. Let x stand for the number of clarinets Orpheus manufactures. Write equations for the total cost of producing x clarinets, and the revenue earned from selling x clarinets.

b. Fill in the table.

x	*Cost*	*Revenue*
0		
50		
100		
150		
200		
250		
300		
350		
400		

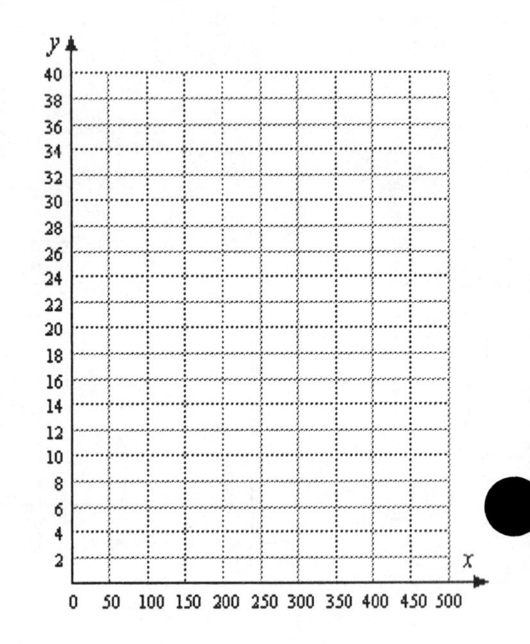

c. Graph both your equations on the grid. (The vertical axis is scaled in thousands of dollars.)

d. How many clarinets must Orpheus sell for their revenue to exceed \$20,000?

e. How many clarinets must Orpheus sell in order to break even?

f. What is their profit if they sell 500 clarinets? 200 clarinets?

g. Illustrate your answers to part (e) on your graph.

32. When sailing upstream in a canal or in a river that has rapids, ships must sometimes negotiate locks to raise them to a higher water level. Suppose your ship is in one of the lower locks, at an elevation of 20 feet. The next lock is at an elevation of 50 feet. Water begins to flow from the higher lock to the lower one, raising your level by one foot per minute, and simultaneously lowering the water level in the next lock by 1.5 feet per minute.

   a. Let t stand for the number of minutes the water has been flowing. Write equations for the water level in each lock after t minutes.

   b. Fill in the table.

t	*Lower Lock Water Level*	*Upper Lock Water Level*
0		
2		
4		
6		
8		
10		

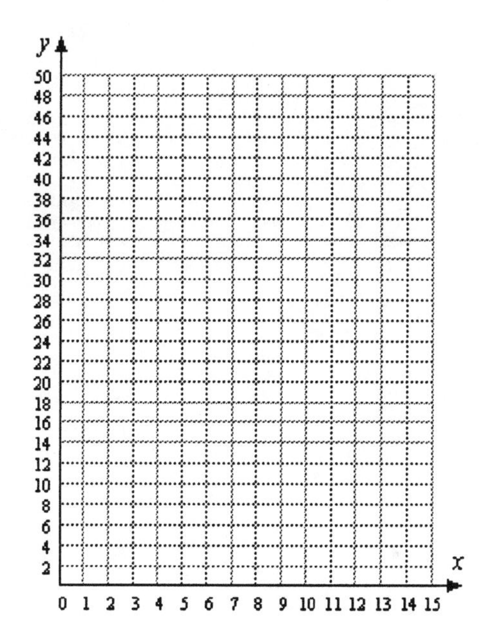

   c. Graph both your equations on the grid.

   d. When will the water levels in the two locks be 10 feet apart?

   e. When will the water level in the two locks be the same?

   f. Write an equation you can use to verify your answer to part (e), and solve it.

Section 3.7 Algebraic Solution of Systems

Exercise 1 Solve the system algebraically:

$$d = 65t$$
$$d = 300 + 50t.$$

(This is the system from Example 2 in Section 3.6.)

Answer:

Exercise 2 Follow the suggested steps to solve the system by substitution:

$$3y - 2x = 3$$
$$x - 2y = -2$$

Answers:

Step 1 Solve the second equation for x in terms of y.

Step 2 Substitute your expression for x into the first equation.

Step 3 Solve the equation you got in Step 2.

Step 4 You now have the solution value for y. Substitute that value into your result from Step 1 to find the solution value for x.

As always, you should check that your solution values satisfy *both* equations in the system.

Exercise 3 Follow the steps to solve the system

$$\begin{aligned} 4x+3y&=7 \\ -9x-3y&=3 \end{aligned}$$

Answers:

Step 1 Add the equations together.

Step 2 Solve the resulting equation for x.

Step 3 Substitute your value for x into either original equation to find the solution value for y.

Step 4 Check that your solution satisfies both original equations.

Exercise 4 Follow the suggested steps to solve the system by elimination.

$$\begin{aligned} 3x&=2y+13 \\ 3y-15&=-7x. \end{aligned}$$

Answers:

Step 1 Write each equation in the form

$$Ax+By=C.$$

Step 2 Eliminate the y-terms: Multiply each equation by an appropriate constant.

Step 3 Add the new equations and solve the result for x.

Step 4 Substitute your value for x into the second equation and solve for y.

Section 3.8 Applications of Systems

Exercise 1 Harvey deposited \$1200 in two accounts. He put \$700 in a savings account that pays 6% annual interest, and the rest in his credit union, which pays 7% annual interest. How much will Harvey's investments earn in two years?

Step 1 Calculate the interest earned by each account.

Savings: $I = Prt =$

Credit union: $I = Prt =$

Step 2 Add the earnings from the two investments.

$$\begin{pmatrix}\text{Total}\\\text{interest}\end{pmatrix} = \begin{pmatrix}\text{Interest from}\\\text{savings}\end{pmatrix} + \begin{pmatrix}\text{Interest from}\\\text{credit union}\end{pmatrix}$$

$$=$$

Exercise 2 You have \$5000 to investfor one year. You want to put part of the money into bonds that pay 7% interest, and the rest of the money into stocks which involve some risk but will pay 12% if successful.

a. Fill in the table.

Amount invested in stocks	*Amount invested in bonds*	*Interest from stocks*	*Interest from bonds*	*Total interest*
\$500				
\$1000				
\$3200				
\$4000				

Now suppose you decide to invest x dollars in the stocks and y dollars in the bonds. Write algebraic expressions for each of the following.

b. Sum of amounts invested: $x + y =$

c. Interest earned on the stocks: $I = Prt =$

Interest earned on the bonds: $I = Prt =$

d. Total interest earned $= \begin{pmatrix}\text{Interest from}\\\text{stocks}\end{pmatrix} + \begin{pmatrix}\text{Interest from}\\\text{bonds}\end{pmatrix}$

$$=$$

Exercise 3 We have two jars of marbles. The first contains 40 marbles, of which 10 are red, and the second contains 60 marbles, of which 30 are red.

a. What percent of the marbles in the first jar are red? $r = \dfrac{P}{W} =$

What percent of the marbles in the second jar are red? $r = \dfrac{P}{W} =$

Now we'll pour both jars of marbles into a larger jar and mix them together.

b. How many marbles total are in the larger jar?

How many red marbles are in the larger jar?

What percent of marbles in the larger jar are red? $r = \dfrac{P}{W} =$

c. Can we add the percents for the first two jars to get the percent red marbles in the mixture?

Exercise 4 In a local city council election your candidate, Justine Honest, ran in a small district with two precincts. Ms. Honest won 30% of the 500 votes cast in Precinct 1 and 70% of the 300 votes cast in Precinct 2. Did Candidate Honest win a majority (more than 50%) of the votes in her district?

Step 1 Fill in the first two rows of the table, using information from the problem and the formula $P = rW$.

	Total votes (W)	*Percent for Honest (r)*	*Votes for Honest (P)*
Precinct 1			
Precinct 2			
Entire district			

Step 2 Add down to complete the first and third columns of the table.

Step 3 Fill in the last entry in the table to answer the question in the problem. Use the formula $P = rW$ again.

Exercise 5 Jerry invested \$2000, part at 4% interest and the remainder at 9%. His yearly income from the 9% investment is \$37 more than his income from the 4% investment. how much did he invest at each rate?

Step 1 Choose variables for the unknown quantities and fill in the table.

	Principal	*Interest Rate*	*Interest*
First investment			
Second investment			

Step 2 Write two equations; one about the principals, and one about the interests.

Principals:

Interests:

Step 3 Solve the system. (Which method seems easiest?)

Exercise 6 Polls conducted by Senator Quagmire's campaign manager show that he can win 60% of the rural vote in his state but only 45% of the urban vote. If 800,000 citizens in urban areas vote, how many voters from rural areas must come to the polls in order for the Senator to win 50% of the vote?

Step 1 Let x represent the number of rural voters and y the total number of voters. Fill in the table.

	Number of Voters (W)	*Percent for Quagmire* (r)	*Number for Quagmire* (P)
Rural			
Urban			
Total			

Step 2 Add down the first and third columns to write a system of equations.

Step 3 Solve your system and answer the question in the problem.

Exercise 7 A river steamer requires 3 hours to travel 24 miles upstream and 2 hours for the return trip downstream. Find the speed of the current and the speed of the steamer in still water.

Step 1 Choose variables.

Speed of the current:

Speed of the steamer:

Step 2 Fill in the table about the steamer.

	Rate	*Time*	*Distance*
Upstream			
Downstream			

Step 3 Write two equations about the steamer.

Step 4 Solve your system and answer the questions in the problem.

Section 3.9 Point-Slope Form

Exercise 1 Compute the slope of the line segment joining A and C in Figure 3.57 in two ways.

a. Find Δy and Δx using the graph.

b. Find Δy and Δx using coordinates.

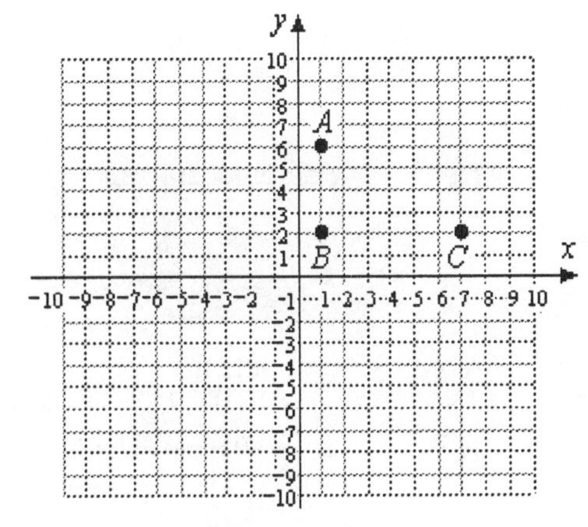

Figure 3.57

Answers:

a. $\Delta y =$ ______ , $\Delta x =$ ______ .

$$m = \frac{\Delta y}{\Delta x} =$$

Draw the line through A and C on Figure 3.57.

b. Write down the coordinates of A and B.

A ____________

B ____________

Write down the coordinates of B and C.

B ____________

C ____________

Compute the directed distance from A to B.

$\Delta y =$ ____________________
final – initial

Compute the directed distance from B to C.

$\Delta x =$ ____________________
final – initial

Compute the slope.

$$m = \frac{\Delta y}{\Delta x} =$$

Do you get the same answers for parts (a) and (b)? You should!

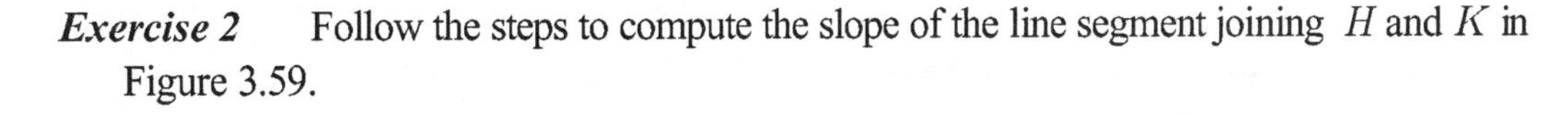

Exercise 2 Follow the steps to compute the slope of the line segment joining H and K in Figure 3.59.

Step 1 Let H be the first point and K the second point. Write down the coordinates of each.

$$H(x_1, y_1) =$$
$$K(x_2, y_2) =$$

Step 2 Fill in the blanks:

$$y_2 = \_\_\_\_\_\_\_\_, \quad y_1 = \_\_\_\_\_\_\_\_$$
$$x_2 = \_\_\_\_\_\_\_\_, \quad x_1 = \_\_\_\_\_\_\_\_$$

Step 3 Compute Δy and Δx.

$\Delta y = y_2 - y_1 =$

$\Delta x = x_2 - x_1 =$

Step 4 Compute the slope.

$$m = \frac{y_2 - y_1}{x_2 - x_1} =$$

Illustrate Δy and Δx on the graph in Figure 3.59. Is your value for the slope reasonable?

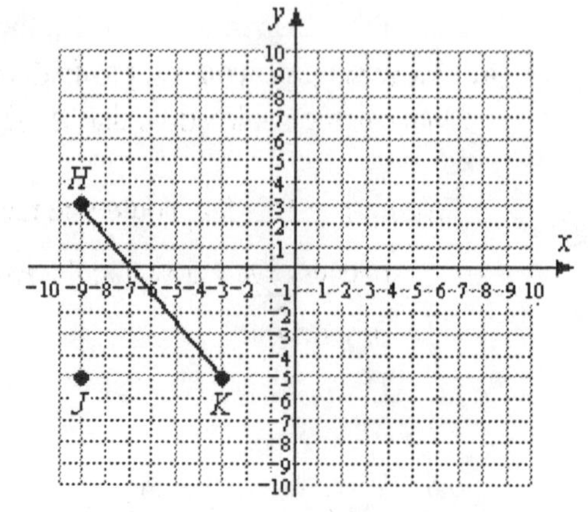

Figure 3.59

Exercise 3 Compute the slope of the line joining the points $(-4,-7)$ and $(2,-3)$.

Answer:

Activity

Suppose we know that our line has slope $m = \frac{3}{2}$, and it passes through the point $(x_1, y_1) = (1,2)$. Follow the steps in the activity below.

1. Plot the point $(1,2)$ on the grid in Figure 3.60.

2. Use Δy and Δx to move to another point on the graph, as you would in the slope-intercept method of graphing.

3. What are the coordinates of the new point, (x_2, y_2)?

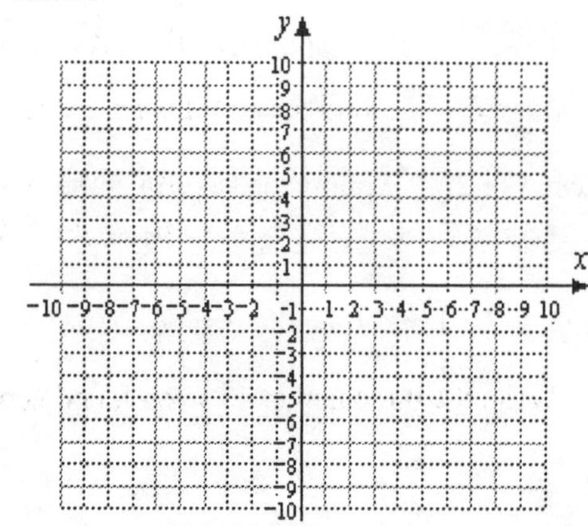

Figure 3.60

Notice that you can find the new coordinates as follows:

$$x_2 = x_1 + \Delta x = \underline{\hspace{3cm}},$$
$$y_2 = y_1 + \Delta y = \underline{\hspace{3cm}}$$

4. Now suppose you want to find a third point on the graph by starting at $(1,2)$ and moving 4 units in the x-direction, that is, $\Delta x = 4$.

***Step 1*:** How far should you move in the y-direction, or what is Δy ? We can find Δy because we know that the *ratio* $\dfrac{\Delta y}{\Delta x}$ must always be $\dfrac{3}{2}$. We can solve the following proportion:

$$\frac{\Delta y}{4} = \frac{3}{2}$$

to find that $\Delta y = 6$. (You should check this.)

***Step 2*:** Use this information to plot the new point on the graph, and to find its new coordinates:

$$x_3 = x_1 + \Delta x = \underline{\hspace{3cm}},$$
$$y_3 = y_1 + \Delta y = \underline{\hspace{3cm}}$$

5. Use the two steps described above to find the coordinates of yet another point on the graph by moving $\Delta x = -6$ from the initial point, $(1,2)$.

Step 1: To find Δy, solve the proportion

$$\frac{\Delta y}{-6} = \frac{3}{2}.$$

Step 2: To find the coordinates of the new point, compute

$$x_4 = x_1 + \Delta x = \underline{\hspace{3cm}},$$
$$y_4 = y_1 + \Delta y = \underline{\hspace{3cm}}$$

6. Write the three new points you found here:

$(x_2, y_2) =$

$(x_3, y_3) =$

$(x_4, y_4) =$

If we use any one of these three points, along with the original point $(1,2)$ to compute the slope, we should get $\frac{3}{2}$. Check that this is true:

a. $\dfrac{\Delta y}{\Delta x} = \dfrac{y_2 - 2}{x_2 - 1} =$

b. $\dfrac{\Delta y}{\Delta x} = \dfrac{y_3 - 2}{x_3 - 1} =$

c. $\dfrac{\Delta y}{\Delta x} = \dfrac{y_4 - 2}{x_4 - 1} =$

In fact, **any point (x, y) on the line must satisfy the following equation:**

$$\frac{y-2}{x-1} = \frac{3}{2},$$

because the slope between any two points is always $\frac{3}{2}$ for this line.

7. Use the equation $\dfrac{y-2}{x-1} = \dfrac{3}{2}$ to decide whether the following points lie on the line:

a. $(-3,-4)$ **b.** $(-5,-3)$ **c.** $(6,4)$

8. Graph the line on Figure 3.60 and check visually whether the three points in part **(6)** are on the line.

Exercise 4 Consider a new line with slope -2 and passing through the point $(1,3)$.

a. Graph the line on the grid.

b. The slope of the line is $m = -2$, so

$$\frac{\Delta y}{\Delta x} = \frac{-2}{1}.$$

The given point is

$$(x_1, y_1) = (1,3).$$

Use the formula

$$\frac{\Delta y}{\Delta x} = \frac{y - y_1}{x - x_1},$$

to write an equation for the line:

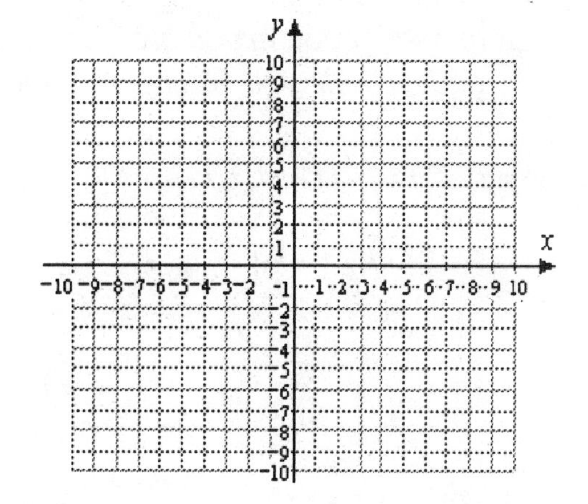

c. Simplify the equation in part (b) to write the equation of the line in the form $y = mx + b$.

What is the y-intercept of the line? Verify your answer against the graph.

Exercise 5 Find an equation for the line of slope $\frac{-1}{2}$ that passes through the point $(-3,-2)$.

Step 1 Use the formula $\frac{y - y_1}{x - x_1} = m.$

Step 2 Cross-multiply to simplify the equation.

Step 3 Solve for y.

Homework 3.9

☐ *In each problem you are given the slope of a line and one point on it. Use the grids to help you find the missing coordinates of the other points on the line.*

13. $m = 2$, $(1, -1)$

a. $(0, ?)$ b. $(?, -7)$

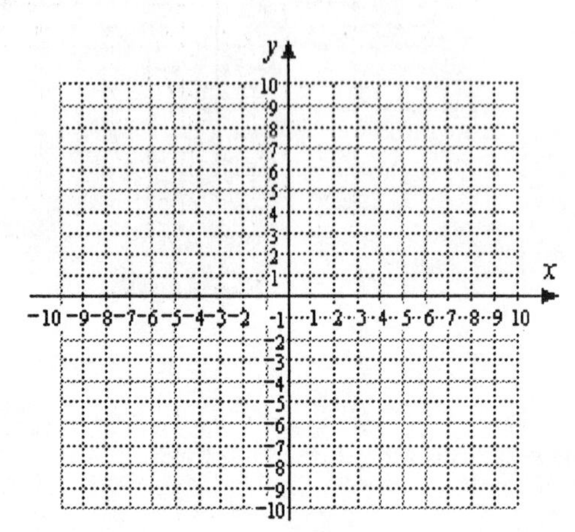

14. $m = -3$, $(-2, 4)$

a. $(-1, ?)$ b. $(?, -5)$

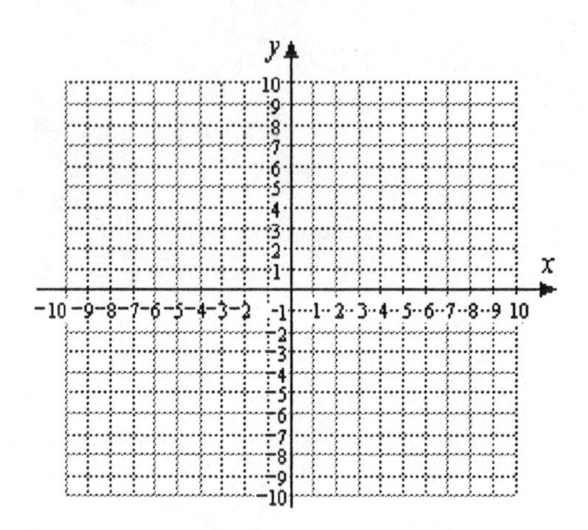

15. $m = \frac{-1}{2}$, $(2, -6)$

a. $(6, ?)$ b. $(?, -3)$

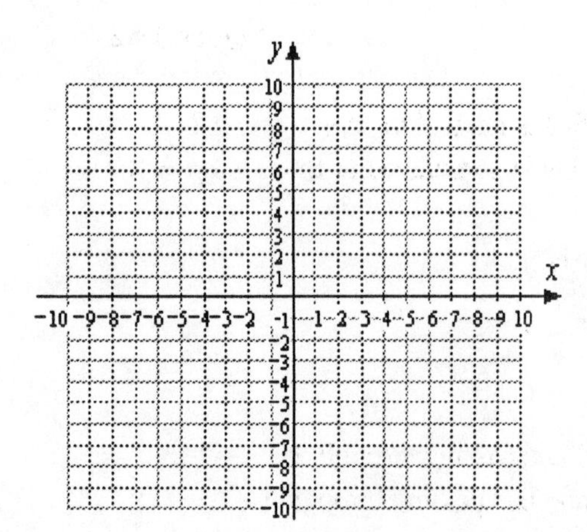

16. $m = \frac{2}{3}$, $(2, -3)$

a. $(8, ?)$ b. $(?, -9)$

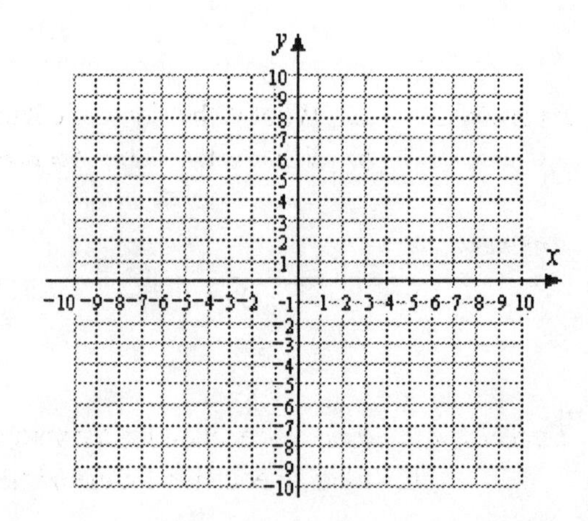

Section 3.10 Using the Point-Slope Form

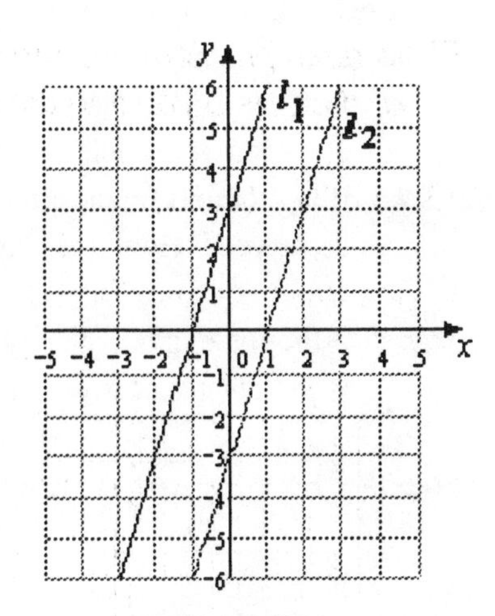

Figure 3.61

Exercise 1 Calculate the slopes of the two parallel lines in Figure 3.61. (We'll call the slope of the first line m_1 and the slope of the second line m_2.)

$$m_1 = \frac{\Delta y}{\Delta x} =$$

$$m_2 = \frac{\Delta y}{\Delta x} =$$

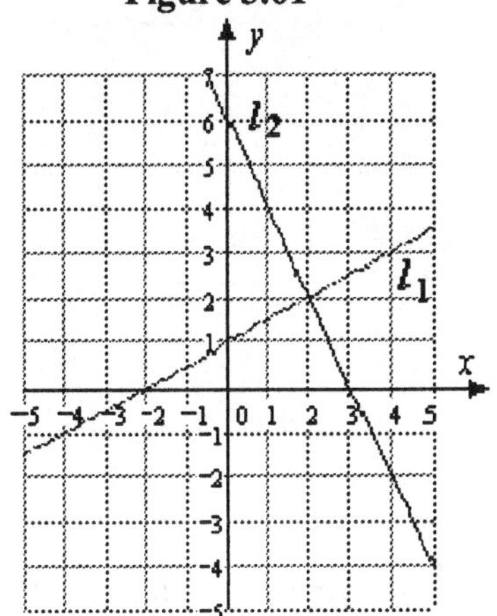

Figure 3.62

Exercise 2 Calculate the slope of the two perpendicular lines in Figure 3.62.

$$m_1 = \frac{\Delta y}{\Delta x} =$$

$$m_2 = \frac{\Delta y}{\Delta x} =$$

Exercise 3 ***a.*** What is the slope of a line that is parallel to $x + 4y = 2$?
b. What is the slope of a line that is perpendicular to $x + 4y = 2$?

Answers a. ***b.***

Exercise 4 a. Find an equation for the vertical line passing through $(-4, -1)$.
b. Find an equation for the horizontal line passing through $(-4, -1)$.

Answers a. ***b.***

Exercise 5 In Section 2.7 we considered some data about the clearing of the world's rain forests. In 1970 there were about 9.8 million square kilometers of rain forest left, and in 1990 there were about 8.2 million square kilometers.

a. Use these data points to find a linear equation for the number of million square kilometers, y, of rain forest left x years after 1950.

b. If we continue to clear the rain forest at the same rate, when will it be completely destroyed? How does your answer compare to the estimate you made in Section 2.7?

Step 1 From the information given, write the coordinates of two points on the line.

Step 2 Find the slope of the line.

Step 3 Use the point-slope formula to find an equation for the line.

Step 4 To answer part (b), find the x-intercept of your line.

Exercise 6 In Section 2.10 we estimated a regression line for data relating shoe size and height. Figure 3.67 shows that regression line for the data.

a. Find (approximate) coordinates of two widely-spaced points on the line. (Since none of the data points lie exactly on the line, you should not choose any of the data points.)

b. Use the point-slope formula to find the equation of the line through your two points.

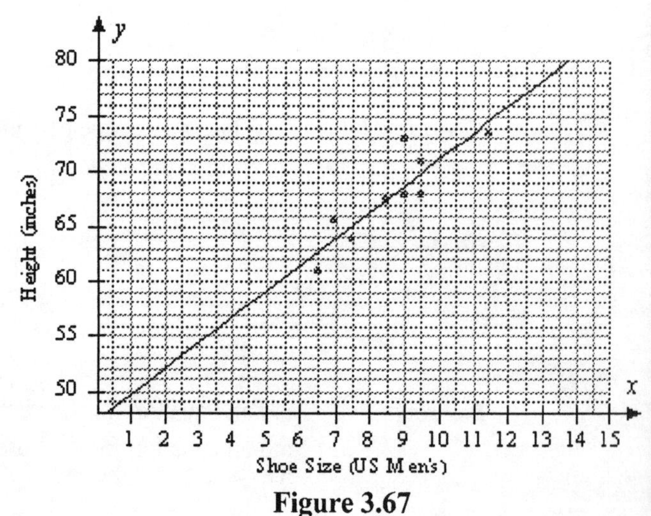

Figure 3.67

c. Use your equation to estimate the height of a person whose shoe size is 8.

d. Use your equation to predict the shoe size of a person who is 70 inches (5' 10") tall.

Homework 3.10

☐ *In Problems 37-40,*

1) *Plot the data on the grid provided.*
2) *Carefully draw a line that fits the data.*
3) *Find the equation of your line, and use the equation to answer each question.*

37. The heights and shadow lengths of several objects were measured at the same time of day. and the results recorded. Plot the data on the grid.

Height (inches)	10	20	30	40	50
Shadow length (inches)	3.8	7.3	11.5	14.6	18.6

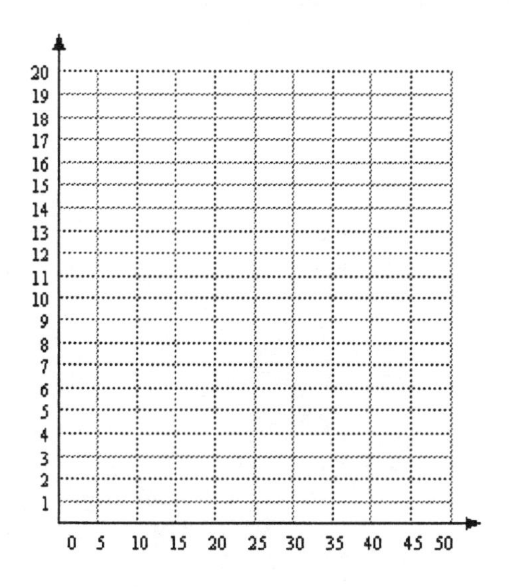

a. Estimate the shadow length of a 17 inch tall object.

b. A maple tree casts an 84 inch long shadow. How tall is the tree?

38. Irene measured the height of a burning candle at different times and recorded the data. Plot the data on the grid.

Time (minutes)	0	5	10	15	20	25	30
Height (cm)	30	29.1	28.7	28.1	27.3	26.6	26.1

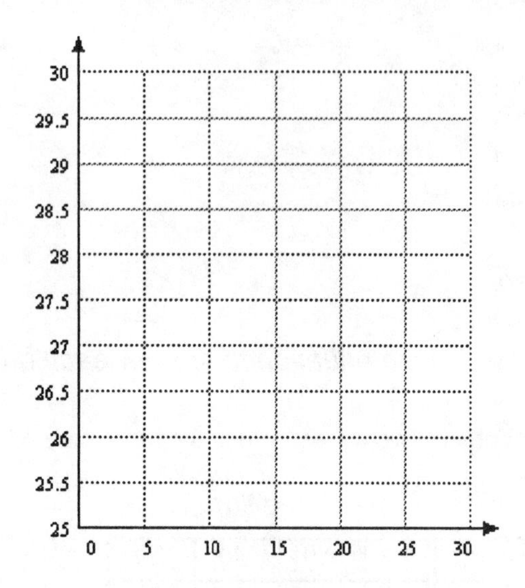

a. What should Irene expect for the height of the candle after 35 minutes?

b. How many minutes should Irene expect the candle to last?

39. As Percy raises the temperature of a gas in a container, the pressure inside the container changes. Plot the data on the grid.

Temperature (° Celsius)	20	40	60	80
Pressure (atmospheres)	1.01	1.08	1.15	1.22

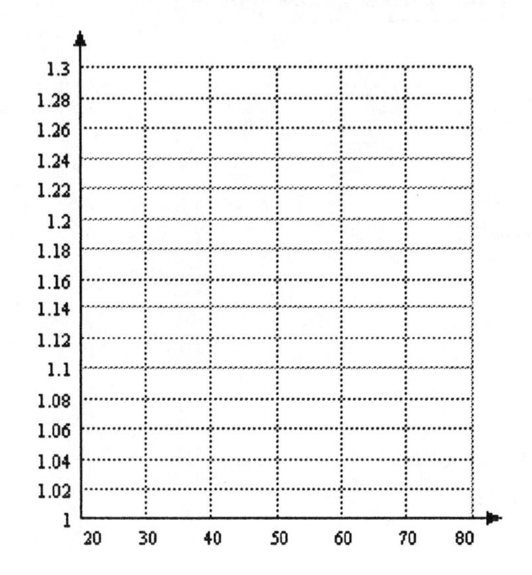

a. Estimate the pressure of the gas at a temperature of 37° Celsius.

b. Estimate the temperature needed to bring the gas to a pressure of 1.5 atmospheres.

c. What is the physical interpretation of the y-intercept of the regression line (if the graph is extended far enough to the right)?

d. What is the physical interpretation of the x-intercept of the regression line (if the graph is extended far enough downwards)?

40. Francine has designed a fuel efficient automobile that runs on a combination of gasoline and battery powered engines. She records how far she can drive on various amounts of gasoline. Plot the data on the grid.

Gasoline (gallons)	0.1	0.2	0.3	0.4
Distance (miles)	6.6	15.5	23.9	33.1

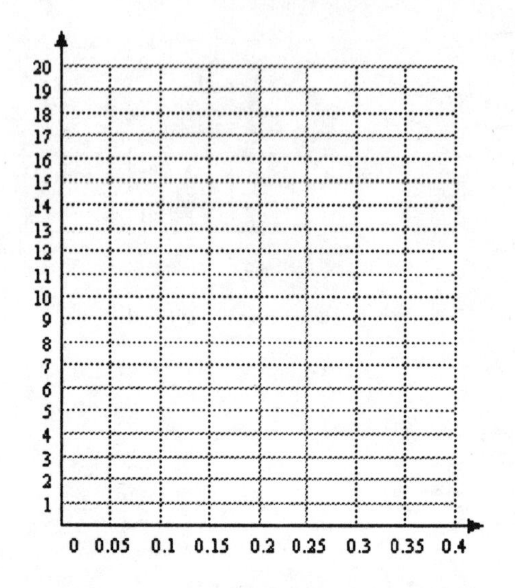

a. Estimate the distance that the automobile can travel on 0.25 gallons of gasoline.

b. Estimate the amount of gasoline needed to drive 50 miles.

c. What is the physical interpretation of the x-intercept of the regression line (if the graph is extended far enough downwards)?

Chapter 3 *Summary and Review*

☐ *Use the graph to find the slope of each line, and illustrate* Δx *and* Δy *on the graph.*

37. $3x+2y=-7$

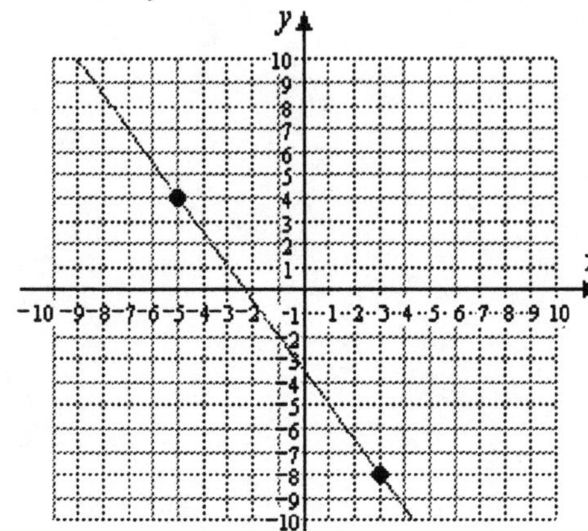

38. $3y=5x$

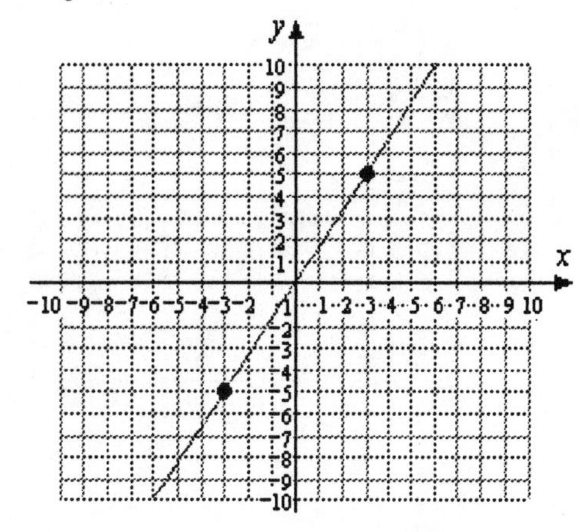

Chapter 4 Quadratic Equations

Section 4.1 Exponents and Formulas

Exercise 1 Compute the following powers. (For part (d), find out how to use your calculator to compute powers.)

a. 6^2 *b.* 3^4 *c.* $\left(\frac{2}{3}\right)^3$ *d.* 1.4^6

Answers: *a.* *b.* *c.* *d.*

Exercise 2 *a.* Find the area of a square whose side is $2\frac{1}{2}$ centimeters long.

b. Find the volume of a cube whose base is the square in part (a). (See Figure 4.2.)

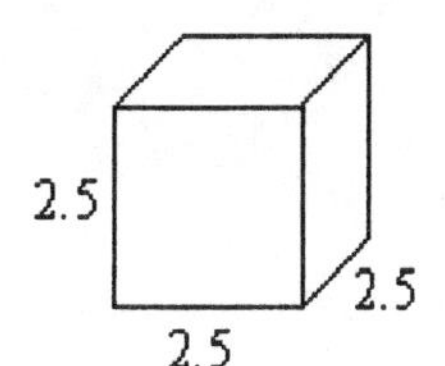

Figure 4.2

Answers: *a.* *b.*

Exercise 3 Compute each power.

a. -4^2 *b.* $(-4)^2$ *c.* $-(4)^2$ *d.* (-4^2)

Answers: *a.* *b.* *c.* *d.*

Exercise 4 Compute each power.

a. $(-2)^1$ *b.* $(-2)^2$ *c.* $(-2)^3$ *d.* $(-2)^4$ *e.* $(-2)^5$

Answers: a. *b.* *c.* *d.* *e.*

f. The sign of the power is ____________ if the exponent is ____________ .

Exercise 5a. Write each expression as a repeated addition or multiplication.

$a^5 =$ ______________________________

$5a =$ ______________________________

b. Evaluate each expression above for $a = 3$.

$a^5 =$ ______________________________

$5a =$ ______________________________

Exercise 6 Compare the expressions by writing them without exponents.

a. $3ab^4$ and $3(ab)^4$ *b.* $a + 5b^2$ and $(a+5b)^2$

Answers: a. $3ab^4 =$ *b.* $a + 5b^2 =$

$3(ab)^4 =$ $(a+5b)^2 =$

Exercise 7 Evaluate each expression for $t = -3$.

a. $-3t^2 - 3$ *b.* $-t^3 - 3t$

Answers: *a.* *b.*

Exercise 8 The surface area of the box in Figure 4.4, whose length is l and whose width and height are both w is given by the formula

$$S = 2w^2 + 4lw.$$

(Do you see how to obtain this formula, using the definition of surface area?)

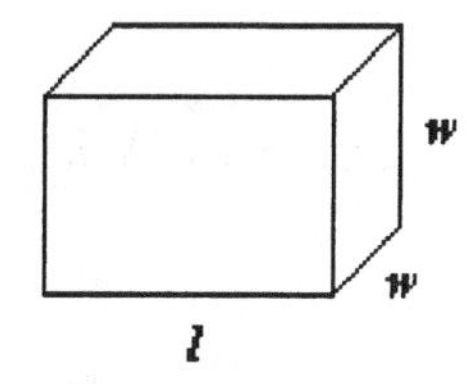

Figure 4.4

a. Francine would like to wrap a package that is 4 feet long and whose width and height are both 1.5 feet. She has 24 square feet of wrapping paper. Is that enough paper?

b. What is the longest box of width and height 1.5 feet that Francine can wrap? (Assume that the paper is just the right shape, and the edges of the paper do not overlap!)

Answers: *a.*

b.

Exercise 9a. Use Example 5 to write a formula for t in terms of A if you invest \$2000 at 5% annual interest.

b. Fill in the table.

Amount (dollars)	2500	3000	5000	6000	7500
Time (years)					

c. Graph your formula. Is the equation linear?

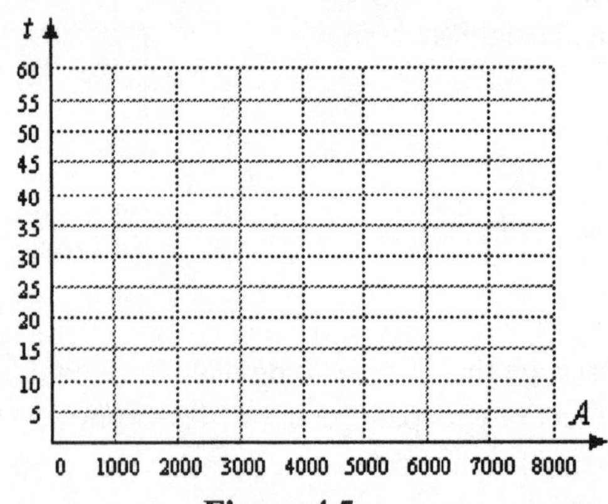

Figure 4.5

Exercise 10 Solve the formula $S = 2w^2 + 4lw$ for l.

Answer:

Section 4.2 Square Roots

Exercise 1 Find two square roots for each number.

a. 225 *b.* $\frac{4}{9}$

Answers: a. *b.*

Exercise 2 Find each square root.

a. $-\sqrt{81}$ *b.* $\pm\sqrt{\frac{64}{121}}$

Answers: *a.* *b.*

Exercise 3 Simplify each radical, if possible.

a. $\sqrt{64}$ *b.* $\sqrt{-64}$ *c.* $-\sqrt{64}$ *d.* $\pm\sqrt{64}$

Answers: *a.* *b.* *c.* *d.*

Exercise 4 Which of the following equations are quadratic?

a. $3x+2x^2=1$ *b.* $4z^2-2z^3+2=0$

c. $36y-16=0$ *d.* $v^2=6v$

Answers:

Exercise 5 Solve $4a^2+25=169$.

Answer:

Exercise 6 Use your calculator to find two approximate solutions to the equation

Answer:

$$2t^2 = 26.$$

Round your answers to three decimal places.

Exercise 7 a. Simplify $\sqrt{6^2 - 4(3)}$.

b. Approximate your answer to part (a) to three decimal places.

c. Explain why the following is *incorrect*:

$$\sqrt{6^2 - 4(3)} \neq \sqrt{6^2} - \sqrt{4(3)} \quad \leftarrow \textit{Incorrect!}$$
$$= \sqrt{36} - \sqrt{12}$$
$$\approx 6 - 3.464 = 2.536$$

Answers: a. ***b.***

c.

Exercise 8 Find each cube root. (Use a calculator for part (d).)

a. $\sqrt[3]{8}$ ***b.*** $\sqrt[3]{-125}$ *c.* $\sqrt[3]{-1}$ *d.* $\sqrt[3]{50}$

Answers *a.* ***b.*** *c.* *d.*

Homework 4.2

☐ *In each exercise, one of the two statements is true for all values of x, and the other is not. By trying some values of x, decide which statement is true.*

39. a. $x + x = 2x$
    b. $x + x = x^2$

x	$x + x$	$2x$	x^2
3			
5			
−4			
−1			

40. a. $x \cdot x = 2x$
    b. $x \cdot x = x^2$

x	$x \cdot x$	$2x$	x^2
4			
6			
−3			
−1			

41. a. $x^2 + x^2 = x^4$
    b. $x^2 + x^2 = 2x^2$

x	$x^2 + x^2$	x^4	$2x^2$
2			
3			
−2			
−1			

42. a. $2x^2 + 3x^2 = 5x^2$
    b. $2x^2 + 3x^2 = 5x^4$

x	$2x^2 + 3x^2$	$5x^2$	$5x^4$
1			
2			
−3			
−2			

43. a. $x + x^2 = x^3$
    b. $x \cdot x^2 = x^3$

x	$x + x^2$	$x \cdot x^2$	x^3
1			
4			
−3			
−1			

44. a. $2x + x^2 = 3x^2$
    b. $2x \cdot x^2 = 2x^3$

x	$2x + x^2$	$2x \cdot x^2$	$3x^2$	$2x^3$
2				
3				
−2				
−1				

Section 4.3 Nonlinear Graphs

Exercise 1: Find the decimal form for each rational number. Does it terminate?

a. $\frac{2}{3}$ *b.* $\frac{5}{2}$ *c.* $\frac{13}{27}$ *d.* $\frac{962}{2000}$

Answers: *a.*

b.

c.

d.

Exercise 2: Give a decimal equivalent for each radical, and identify it as rational or irrational. If necessary, round your answers to three decimal places.

$\sqrt{1} =$ ____________ $\sqrt{6} =$ ____________

$\sqrt{2} =$ ____________ $\sqrt{7} =$ ____________

$\sqrt{3} =$ ____________ $\sqrt{8} =$ ____________

$\sqrt{4} =$ ____________ $\sqrt{9} =$ ____________

$\sqrt{5} =$ ____________ $\sqrt{10} =$ ____________

☐ *For each of the problems below:*
   a. *Fill in the table of values,*
   b. *Graph the equation,*
   c. *Answer the questions.*

1. At the Custom Pizza shop you can buy their special smoked chicken pizza in any size you like. The cost, C, of the pizza in dollars is given by the equation

$$C = \frac{1}{4}r^2,$$

where r is the radius of the pizza. (Use Figure 4.9.)

r	C
0	
1	
2	
4	
6	
9	
10	
14	

a. How much does a pizza of radius 3 inches cost? Locate this point on your graph.

b. Use your graph to find out how big a pizza you can buy for $16.

c. If you can spend at most $36 on pizza, what sizes can you buy? Mark all of these on the horizontal axis.

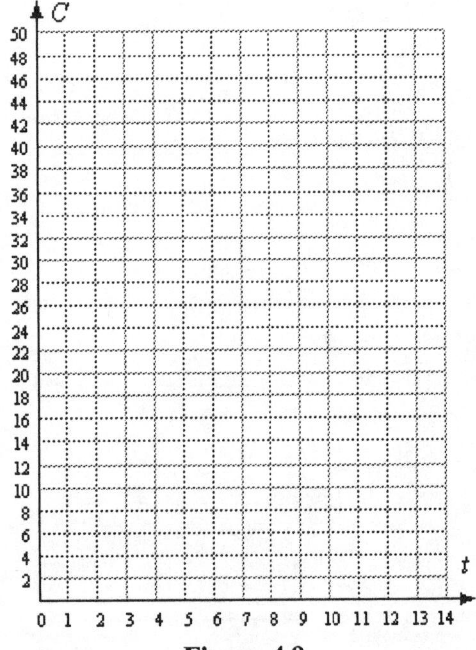

Figure 4.9

2. While hiking, Delbert drops a stone off the edge of a 400-foot cliff. After falling for t seconds, the height h of the stone above the base of the cliff is given in feet by

$$h = 400 - 16t^2.$$

t	h
0	
1	
2	
3	
4	
5	

a. What is the height of the stone 3 seconds after being dropped? Locate this point on your graph in Figure 4.10.

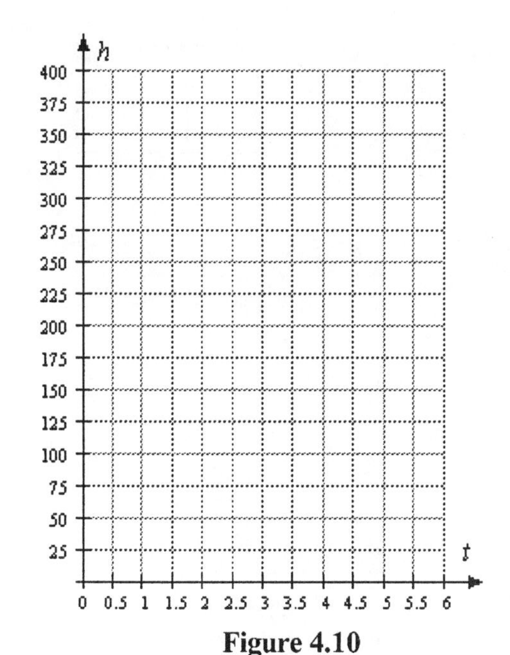

Figure 4.10

b. Use your graph to find out how long will it take the stone to hit the ground.

c. On the horizontal axis, mark all of the times when the stone is above 300 feet high.

3. The area of a circle is given by the formula

$$A = \pi r^2,$$

where r is the radius of the circle.

a. Rewrite the formula using an approximation for π rounded to two decimal places.

r	A
0	
0.5	
1	
1.5	
2	
2.5	
3	
3.5	
4	

b. Fill in the table of values, rounding your answers to one decimal place, then graph the equation on the grid in Figure 4.11.

c. Use your graph to estimate the area of a circle whose radius is 2.25 inches. Then find the area algebraically, using the formula.

d. Use your graph to estimate the radius of a circle whose area is 30 square inches. Then find the radius algebraically by using the formula.

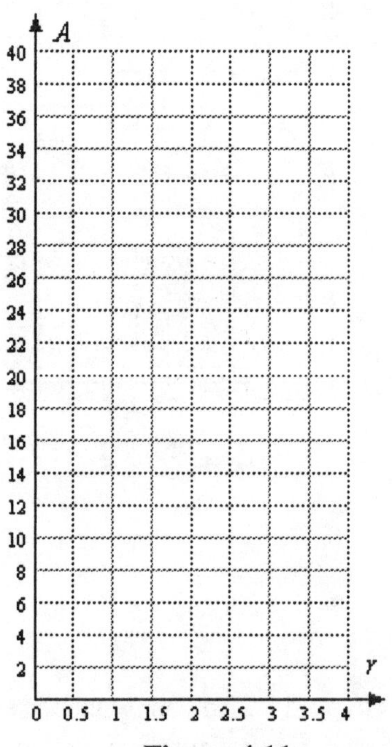

Figure 4.11

4. The volume of a sphere of radius r is given by the formula

$$V = \frac{4}{3}\pi r^3.$$

a. Rewrite the formula using a decimal approximation for $\frac{4}{3}\pi$ rounded to two decimal places.

b. Fill in the table of values, rounding your answers to one decimal place, then graph the equation on the grid in Figure 4.12.

r	0	0.5	1	1.5	2	2.5	3	3.5	4
V									

c. Use your graph to estimate the volume of a sphere whose radius is 2.25 inches. Then find the area algebraically, using the formula.

d. Use your graph to estimate the radius of a sphere whose volume is 14 cubic inches. Then find the radius algebraically by using the formula.

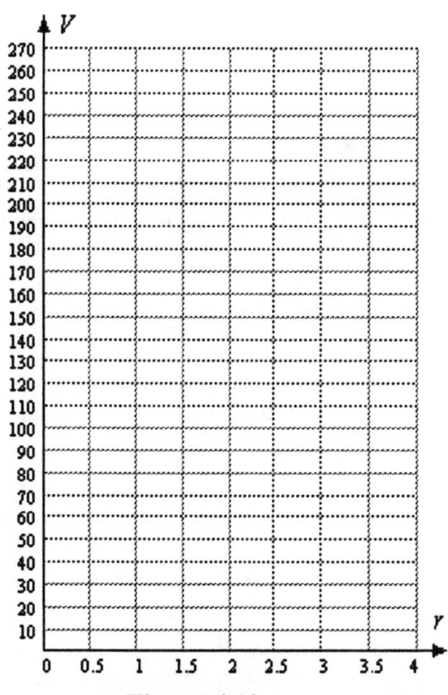

Figure 4.12

5. If you are flying in an airplane at an altitude of h feet, on a clear day you can see for a distance of d miles, where d is given by

$$d = 1.22\sqrt{h}\,.$$

h	d
2000	
5000	
10,000	
15,000	
20,000	
25,000	
30,000	

a. Fill in the table of values, then graph the equation on the grid in Figure 4.13.

b. How far can you see from an altitude of 10,000 feet? Locate this point on your graph.

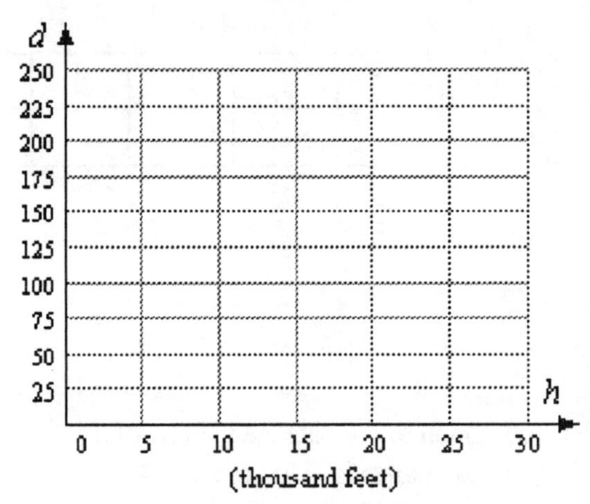

Figure 4.13

c. How high do you have to be to see 100 miles? Locate this point on your graph.

d. On the horizontal axis, mark all altitudes from which you can see further than 175 miles.

6. If an object falls from a height of h meters, its velocity v when it strikes the ground is given in meters per second by

$$v = 4.4\sqrt{h}\ .$$

a. Fill in the table of values, then graph the equation on the grid in Figure 4.14.

h	50	100	150	200	250	300	350	400
v								

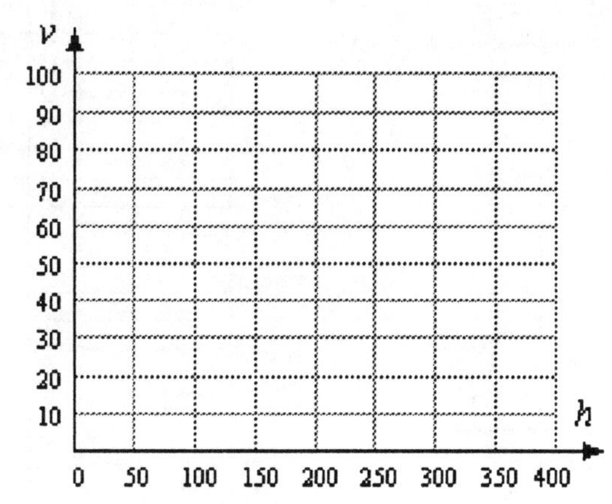

Figure 4.14

b. If a penny falls off the Washington monument, 170 meters high, what will its velocity be when it hits the ground? Locate this point on your graph.

c. A rock dropped from the Royal Gorge bridge in Colorado will hit the water below at a velocity of 80 meters per second. How high is the bridge? Locate this point on your graph.

d. On the horizontal axis, mark all heights at which the velocity of the penny is less than 70 meters per second.

Homework 4.3

☐ *For problems 23-30, first review Section 2.6 on solving inequalities graphically. Graph each equation by plotting points. Then use your graph to solve the equations and inequalities given. (You may have to estimate the solutions.)*

23. $y = x^2$

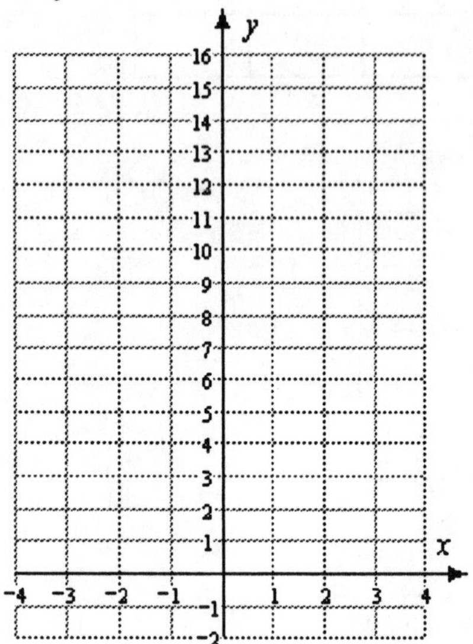

a. $x^2 = 12$ b. $x^2 \leq 6$

24. $y = x^3$

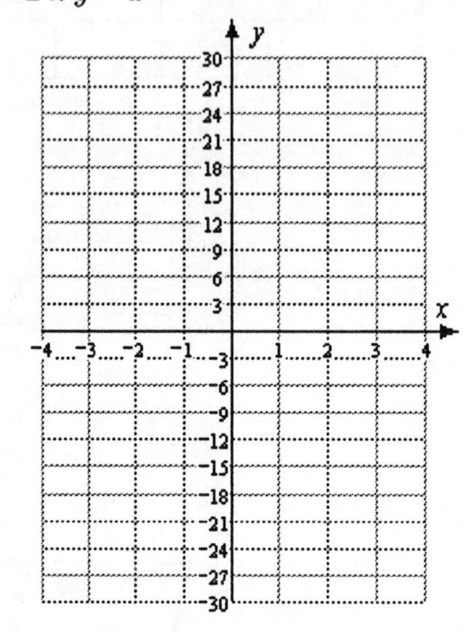

a. $x^3 = -15$ b. $x^3 > -2$

25. $y = \sqrt{x}$

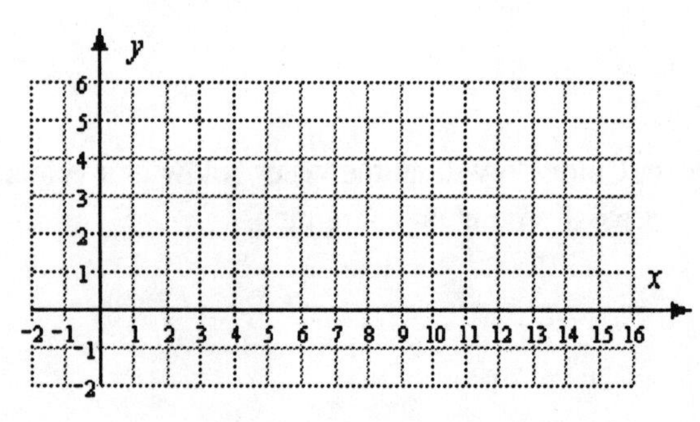

a. $\sqrt{x} = 2.5$ b. $1 < \sqrt{x} \leq 3$

26. $y = \sqrt[3]{x}$

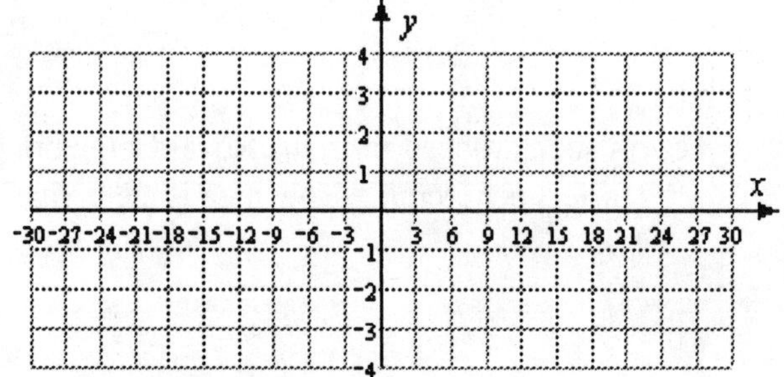

a. $\sqrt[3]{x} = -2.5$ b. $2 \geq \sqrt[3]{x} \geq -2$

27. $y = 4 - x^2$

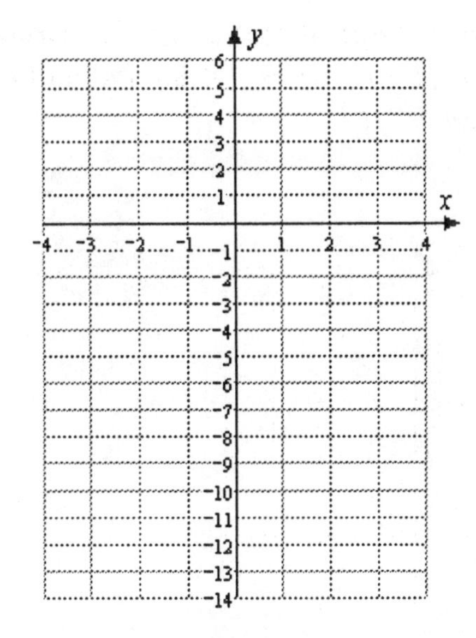

a. $4 - x^2 = -5$
b. $4 - x^2 < 0$

28. $y = (x+3)^2$

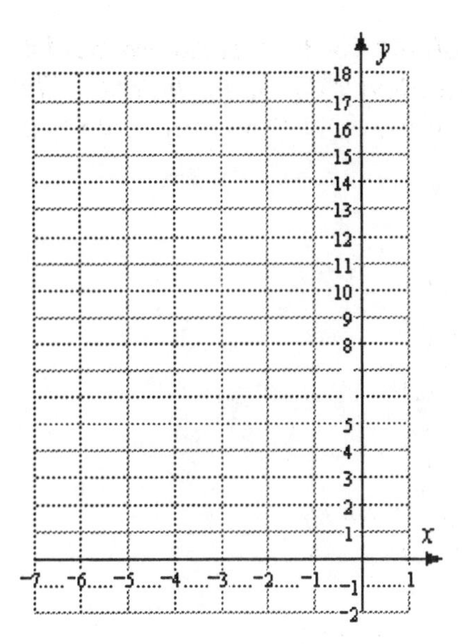

a. $(x+3)^2 = -1$
b. $(x+3)^2 \leq 8$

29. $y = \sqrt{x+2}$

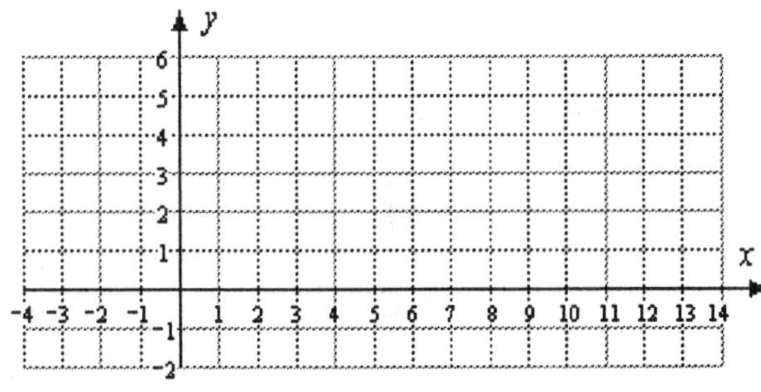

a. $\sqrt{x+2} = 3$
b. $\sqrt{x+2} > 2$

30. $y = \sqrt{2-x}$

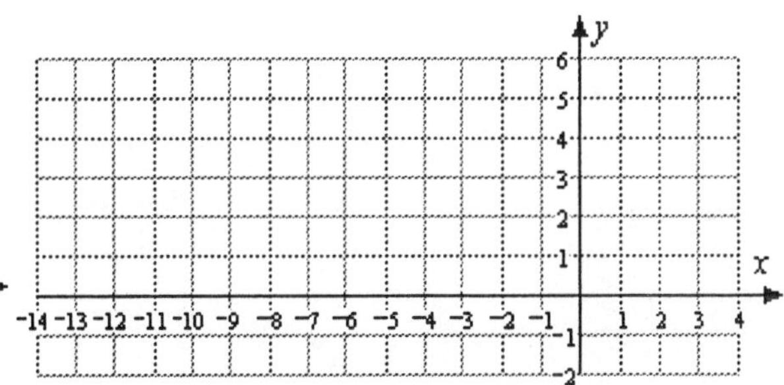

a. $\sqrt{2-x} = 3$
b. $\sqrt{2-x} > 2$

Section 4.4 Pythagorean Theorem

Exercise 1 ***a.*** Sketch the situation described in the story. Show Delbert's camp, the canyon and the bridge, Delbert's actual route and the route he should have taken.

b. Without doing any calculations, guess how far Delbert must ride on the dirt road to reach the bridge.

Activity

In this activity we investigate some properties of right triangles. The grid provided in the back of your workbook is measured in centimeters. Use the grid and follow these steps:

Step 1 First, measure carefully and cut out a strip of paper or cardboard between 12 and 24 centimeters long. (It will be easier if you choose an integer value for the length of the strip.)

Step 2 Use your strip as the hypotenuse of a right triangle on the grid, with the legs on the axes.

Step 3 Record the lengths of the legs (and the hypotenuse) in the first three columns of Table 4.1. (We'll get to the other columns in a minute.)

Step 4 Try several different triangles, and record the lengths of the legs.

Table 4.1

Hypotenuse (H)	*First Leg* (FL)	*Second Leg* (SL)	$(FL)^2$	$(SL)^2$	$(FL)^2 + (SL)^2$	$(H)^2$

Repeat the exercise above with a hypotenuse strip of a different length, and record your data in the first three columns of Table 4.2.

Table 4.2

Hypotenuse (H)	*First Leg* (FL)	*Second Leg* (SL)	$(FL)^2$	$(SL)^2$	$(FL)^2 + (SL)^2$	$(H)^2$

Step 5 The last four columns in each table are for recording some calculations. You should be able to interpret what goes in each column:

$(FL)^2$:	Square the length of the first leg.
$(SL)^2$:	Square the length of the second leg.
$(FL)^2 + (SL)^2$:	Add the entries in the previous two columns.
$(H)^2$:	Square the length of the hypotenuse.

Step 6 Compare the entries in the last two columns of each table. Do you notice any pattern? Summarize your observations as a conjecture in the space below.

Conjecture:

Exercise 2 Use the grid to model the sketch you made in Exercise 1 for Delbert's trail bike trip. (How long should you cut the hypotenuse strip?) From the grid, read the distance that Delbert must travel along the dirt road to get to the bridge.

Exercise 3 Use the Pythagorean theorem to write an equation about your sketch in Exercise 1. Solve the equation to find out how far Delbert must ride on the dirt road before reaching the bridge. How does your answer compare with your estimate in Exercise 2?

Grid for Section 4.4: The Pythagorean Theorem

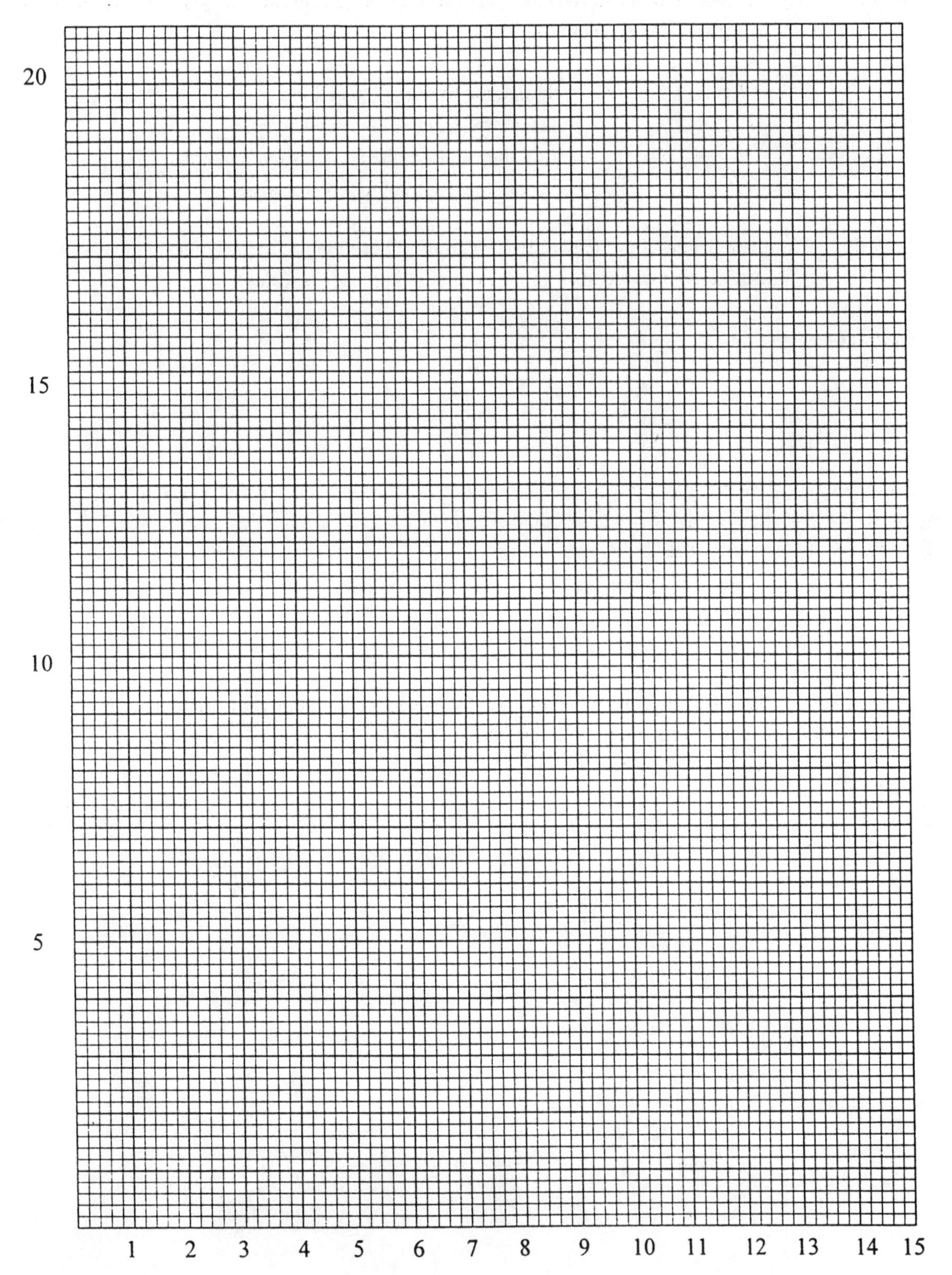

Homework 4.4

☐ *In Exercises 25-28, decide whether the two expressions are equivalent by evaluating them for several values of the variables.*

25. Does $(a+b)^2 = a^2 + b^2$?

a	b	$a+b$	$(a+b)^2$	a^2	b^2	a^2+b^2
2	3					
3	4					
1	5					
−2	6					

26. Does $\sqrt{a^2+b^2} = a+b$?

a	b	$a+b$	a^2	b^2	a^2+b^2	$\sqrt{a^2+b^2}$
3	4					
2	5					
1	6					
−2	−3					

27. Does $\sqrt{a+b} = \sqrt{a} + \sqrt{b}$?

a	b	$a+b$	$\sqrt{a+b}$	$\sqrt{a}$	$\sqrt{b}$	$\sqrt{a}+\sqrt{b}$
2	7					
4	9					
1	5					
9	16					

28. Does $(\sqrt{a}+\sqrt{b})^2 = a+b$?

a	b	$a+b$	$\sqrt{a}$	$\sqrt{b}$	$\sqrt{a}+\sqrt{b}$	$(\sqrt{a}+\sqrt{b})^2$
4	9					
1	4					
3	5					
6	10					

Section 4.5 Products of Binomials

Exercise 1

Find the area of each rectangle.

Use the distributive law to find the product.

a.

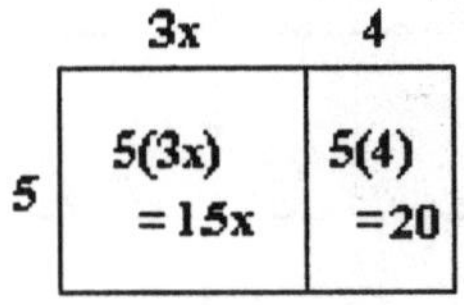

	3x	4
5	5(3x) = 15x	5(4) = 20

Area = 15x + 20

a. $5(3x+4) = 5(3x) + 5(4)$

$= 15x + 20$

b.

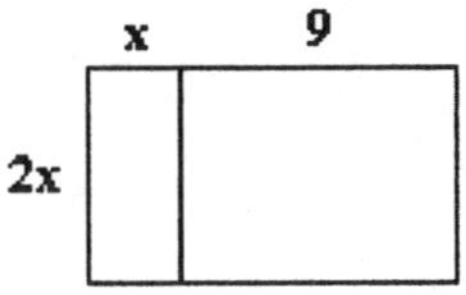

	x	9
2x		

b. $2x(x+9) =$

c.

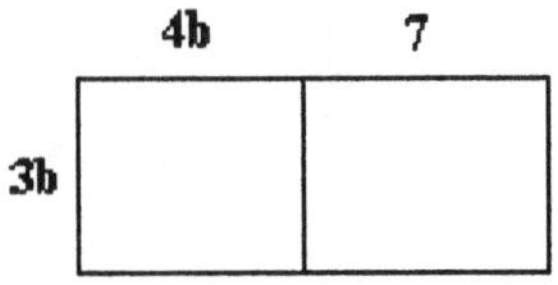

	4b	7
3b		

c. $3b(4b+7) =$

Exercise 2 Use the distributive law to find the products.

a. $2a(6a-5)$

b. $-4v(2v-3)$

c. $-5x(x^2-3x+2)$

d. $-3xy(4x^2-2xy+2y)$

Answers: ***a.***

b.

c.

d.

Exercise 3 Write the area of each rectangle in two different ways: as the sum of four smaller areas, and as one large rectangle, using the formula $Area = length \times width.$

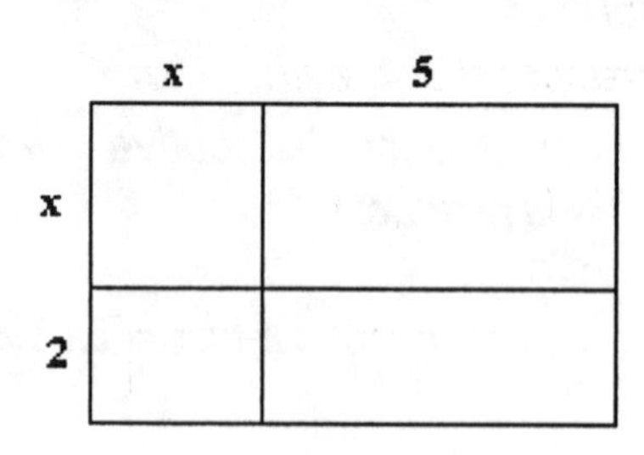

2x 4

x

3

Answers: ***a.*** ***b.***

Exercise 4 ***a.*** Use a rectangle to represent the product $(3x-2)(x-5)$.
b. Write the product as a quadratic trinomial.

Answers ***a.*** ***b.***

Exercise 5 Use the rectangle diagram to help you find the linear term in the product

$$(x-6)(2x+3).$$

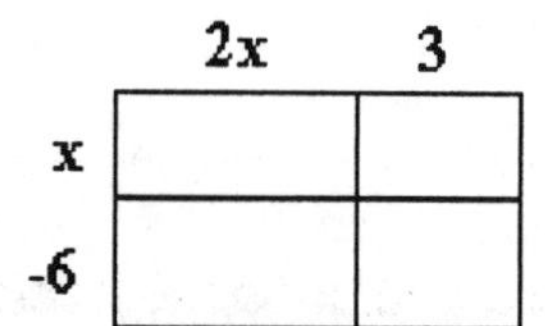

Answer:

Exercise 6 Write each product as a quadratic trinomial in two variables.

a. $(3a-5b)(3a-b)$ ***b.*** $(x+4y)^2$

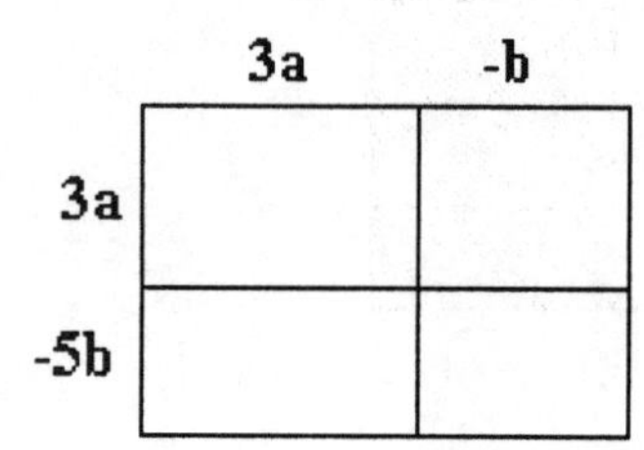

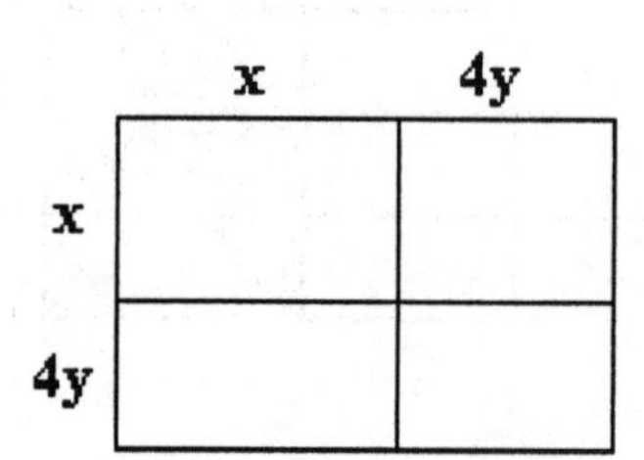

Answers:

a. ***b.***

Homework 4.5

☐ *a. Find the linear term in each product.*
b. Shade the sub-rectangles that correspond to the linear term.

43. $(x+6)(x-9)$

44. $(x-5)(x+3)$

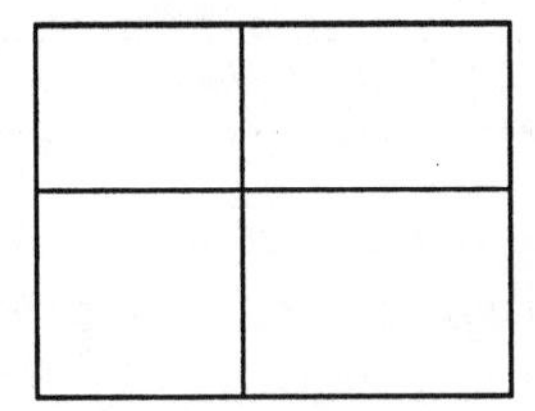

45. $(2x-5)(x+4)$

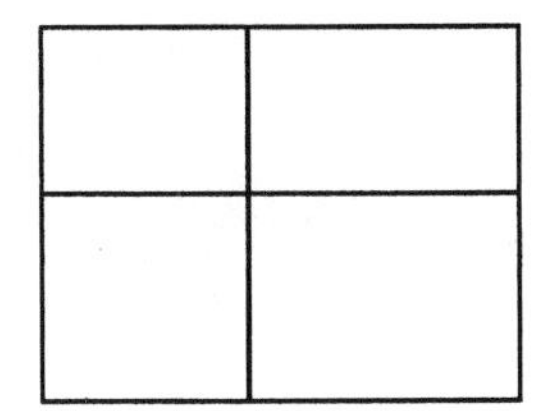

46. $(3x-2)(2x+1)$

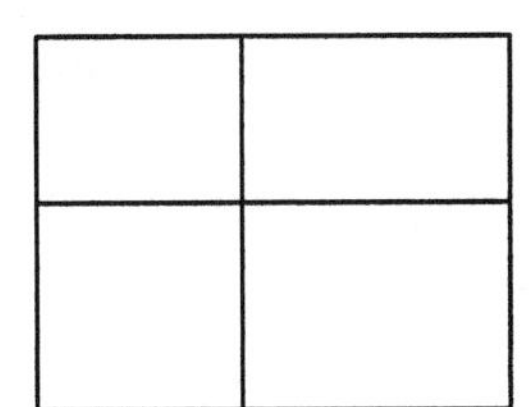

☐ *Use evaluation to decide whether the two expressions are equivalent.*

65. $(a-b)^2$ and a^2-b^2

a	b	$a-b$	$(a-b)^2$	a^2	b^2	a^2-b^2
5	3					
2	6					
-4	-3					

66. $(a+b)^2$ and a^2+b^2

a	b	$a+b$	$(a+b)^2$	a^2	b^2	a^2+b^2
2	3					
-2	-3					
2	-3					

☐ *Write the area of each square in two different ways:*
a. as the sum of four smaller areas,
b. as one large square, using the formula $Area = (length)^2$.

67.

68.

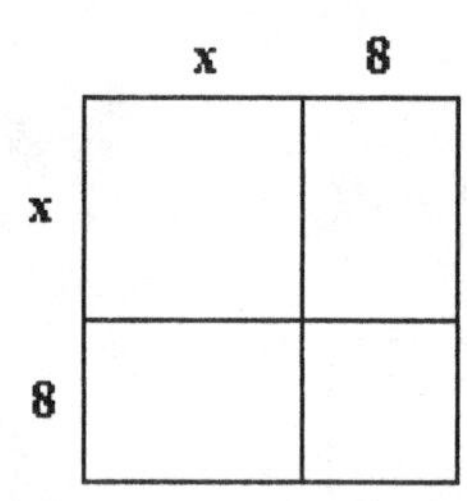

69.

70.

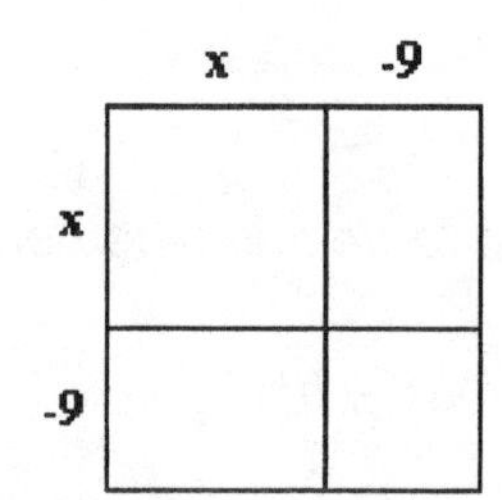

Midchapter Review

☐ *Graph the indicated equation by plotting points, then use your graph to estimate solutions to the equations in (a), (b), and (c).*

37. $y = x^3$

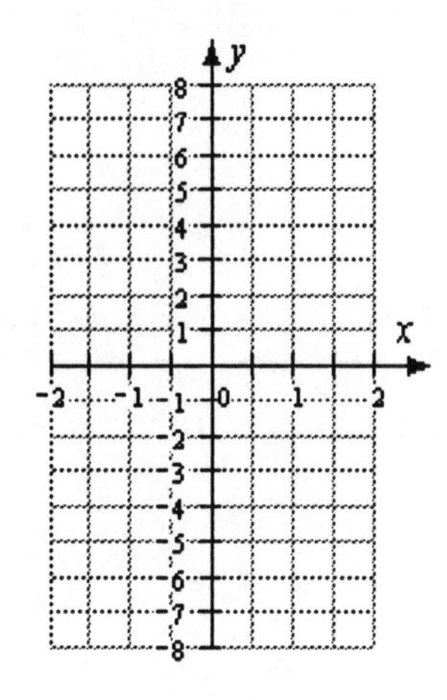

a. $x^3 = 2$

b. $x^3 = -1.4$

c. $x = \sqrt[3]{4}$

38. $y = x^2$

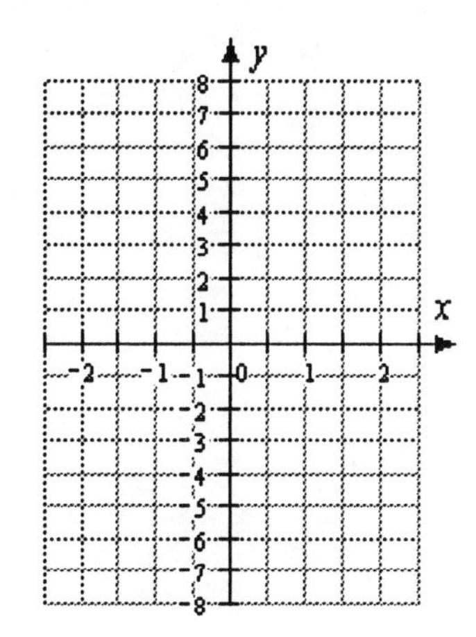

a. $x^2 = 5$

b. $x^3 = -3$

c. $x = -\sqrt{2}$

Section 4.6 Graphing Quadratic Equations

Exercise 1a. Use the formula for revenue to writean equation for Raingear's revenue in terms of p:

$$R =$$

Simplify your equation by applying the distributive law:

$$R =$$

b. Write an equation for part (b) of Problem 1: How much should Raingear charge for each umbrella if they would like to make $\$1000$ monthly revenue?

Answer:

Exercise 2a. Use the table to look for a solution to part (b) of Problem 1.

Price per umbrella p	*No. of umbrellas sold* $150-5p$	*Revenue* $R=$
2		
4		
6		
8		
10		
12		
14		

b. Will the revenue continue to increase as the price per umbrella increases? Why or why not?

Answers: ***a.***

b.

Skills Review

☐ a. *Solve the equation.*
   b. *Write the equation in the form* $ax + b = 0$.
   c. *Graph the equation* $y = ax + b$.
   d. *Find the* x*-intercept of your graph. Compare with your answer to part (a).*

1. $2x+5=11$

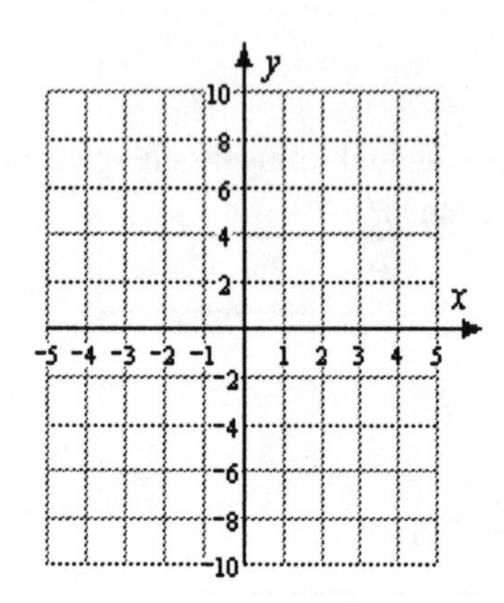

2. $2x-3=5x+9$

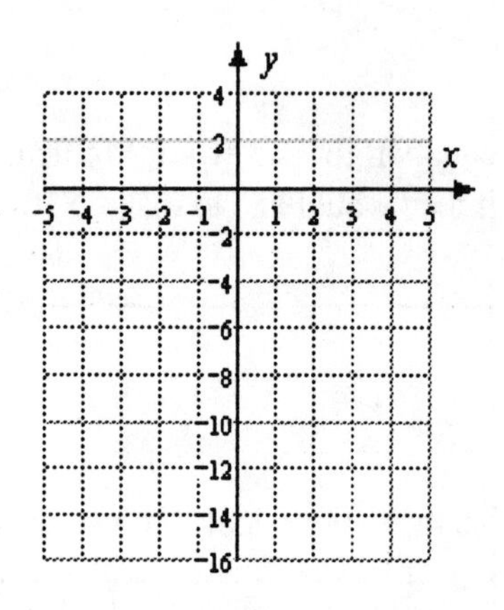

3. $0.7x+0.2(100-x)=0.3(100)$

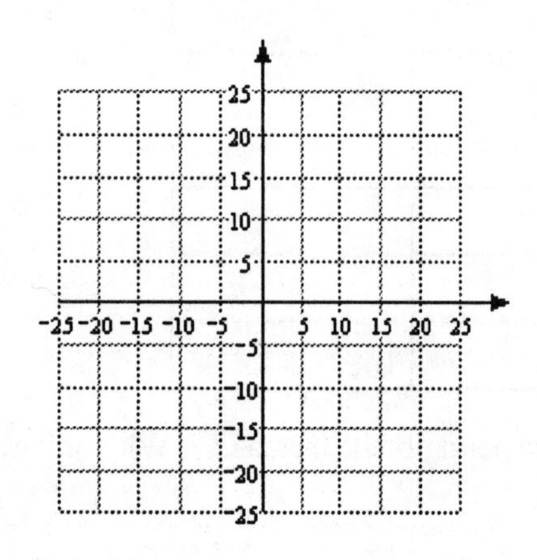

4. $4(7-4x)=-2(6x-5)-6$

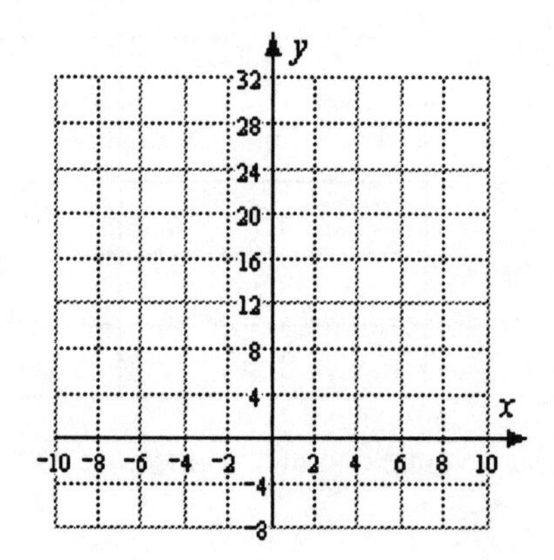

Activity

For Exercise 1 in the Reading assignment you wrote a quadratic equation about the total revenue Raingear will earn from selling umbrellas. A good way to study such problems is to consider the graph of the equation.

a. Extend your table of values from Exercise 2 by completing the table below.

Price per umbrella p	*Number of umbrellas sold* $150-5p$	*Revenue* $R=$
11		
13		
15		
17		
18		
20		
22		
24		
28		
30		

b. On the grid in Figure 4.28, use your tables from Exercises 2 and 3 to draw a smooth graph of the equation for revenue, R, in terms of p.

c. Locate the point on your graph that corresponds to a monthly revenue of $1000.

d. Find another point on the graph that corresponds to $1000 monthly revenue. What price per umbrella produces this revenue? How many umbrellas will be sold at this price?

Price per umbrella:

Number of umbrellas:

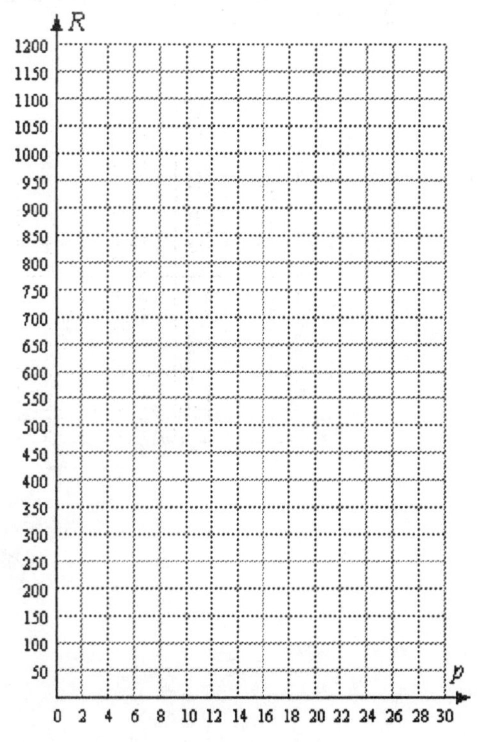

Figure 4.28

e. What is the maximum monthly revenue that Raingear can earn from umbrellas?

Maximum revenue:

f. What price per umbrella should they charge in order to earn the maximum revenue?

Price:

Exercise 3 Complete the table of values and graph each of the quadratic equations below. Notice the similarities and differences among the graphs.

a. $y = x^2 + 4$

x	-3	-2	-1	0	1	2	3
y							

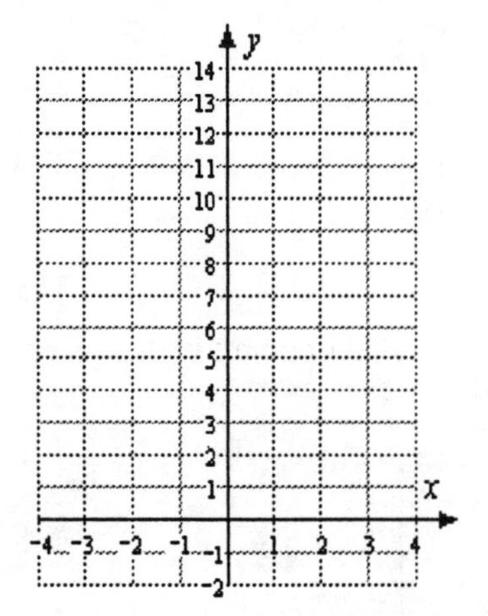

b. $y = x^2 - 4$

x	-3	-2	-1	0	1	2	3
y							

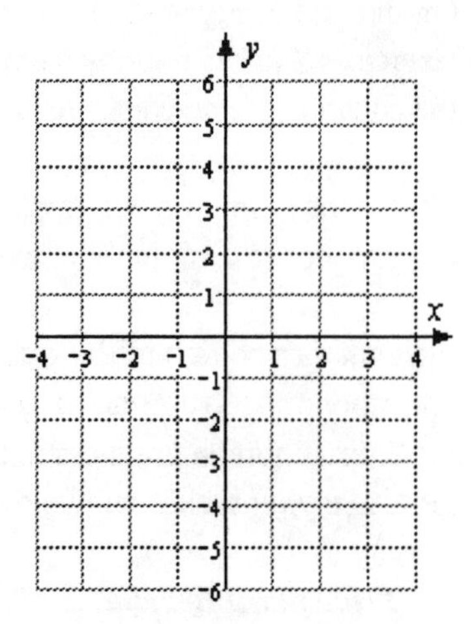

c. $y = 4 - x^2$

x	-3	-2	-1	0	1	2	3
y							

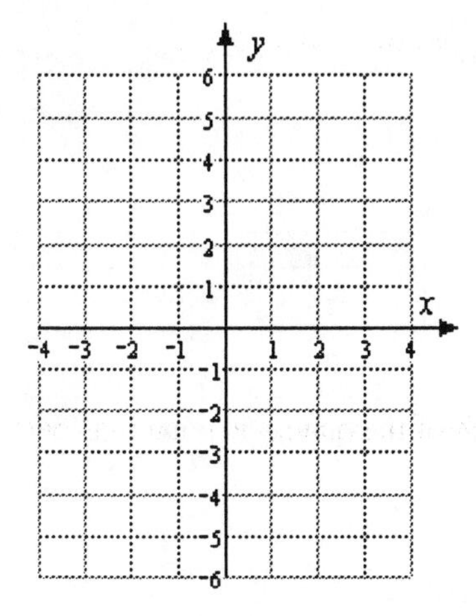

d. $y = (x-4)^2$

x	1	2	3	4	5	6	7
y							

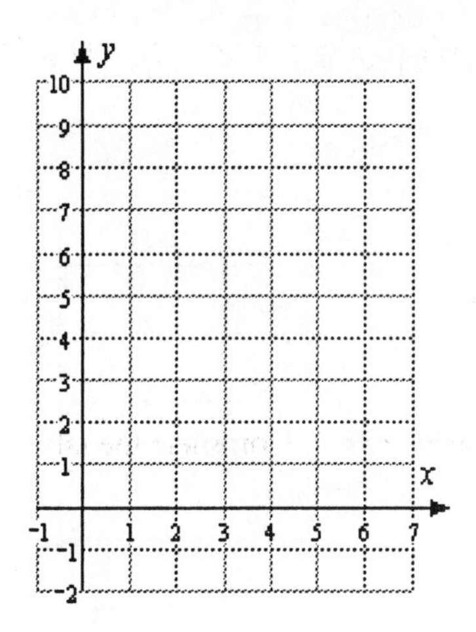

Exercise 4 Find the vertex of each of the parabolas in Exercise 3.

Answers:

Exercise 5a. For each of the graphs in Exercise 3, locate the point or points on the graph (if they exist) that correspond to $y = 0$.

b. Explain how to use a graph to solve the equation $ax^2 + bx + c = 0$.

Answers ***a.***

b.

Problem 2 Delbert is standing at the edge of a 360-foot cliff. He throws his algebra book upwards off the cliff with a velocity of 36 feet per second. The height of his book above the ground (at the base of the cliff) after t seconds is given by the formula

$$h = -16t^2 + 36t + 360,$$

where h is in feet.

a. Fill in the table, and then graph the equation. (Use your calculator to evaluate the expression for h.)

t	h
0	
0.5	
1	
1.5	
2	
2.5	
3	
4	
5	
6	

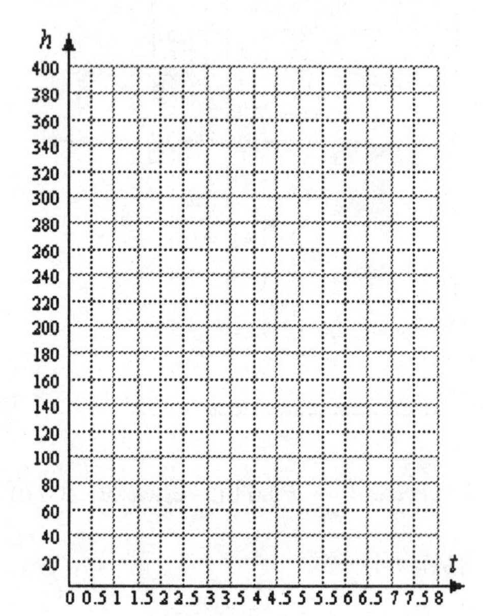

Figure 4.30

b. What is the height of the book after 1 second? After 3 seconds?

c. Use your graph to find the approximate time when the book is 200 feet high.

d. When does the book hit the ground?

e. Write an algebraic equation that we could solve in order to find out when the book hits the ground.

f. Verify that your answer to part (d) is a solution for your equation in part (e).

Homework 4.6

☐ *Make a table of values and graph each pair of parabolas on the grid provided.*

1. a. $y = x^2 + 1$
   b. $y = x^2 - 3$

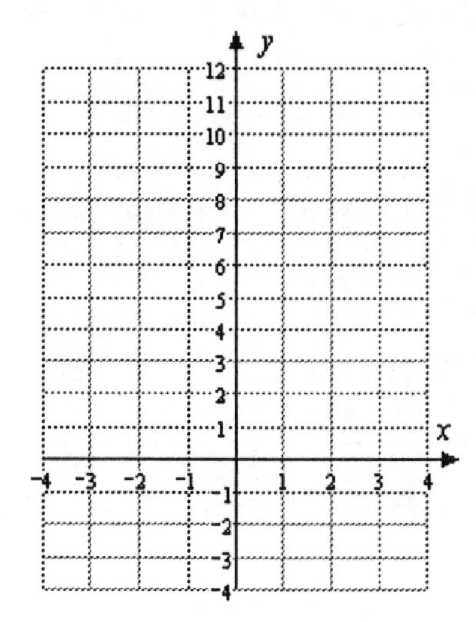

2. a. $y = x^2 - 1$
   b. $y = x^2 + 2$

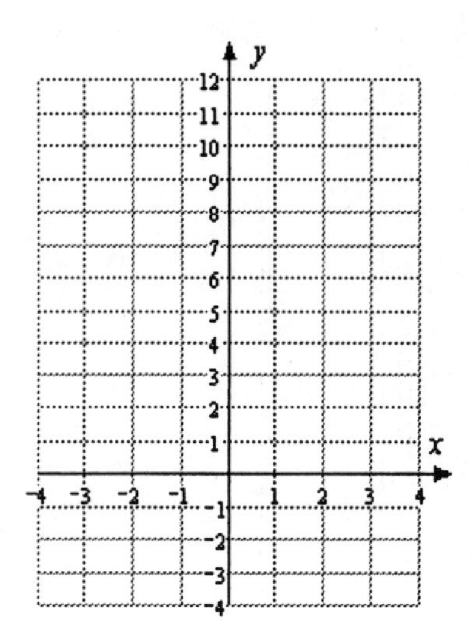

3. a. $y = -x^2 - 2$
   b. $y = 5 - x^2$

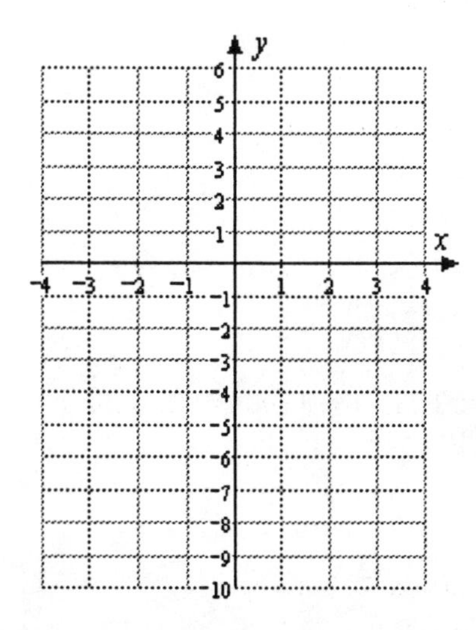

4. a. $y = -x^2 + 3$
   b. $y = 9 - x^2$

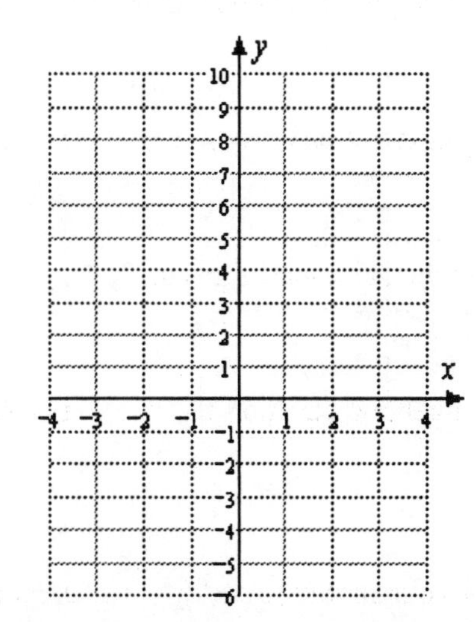

☐ *Make a table of values and graph each pair of parabolas on the grid provided.*

7. a. $y = (x+2)^2$
   b. $y = (x-1)^2$

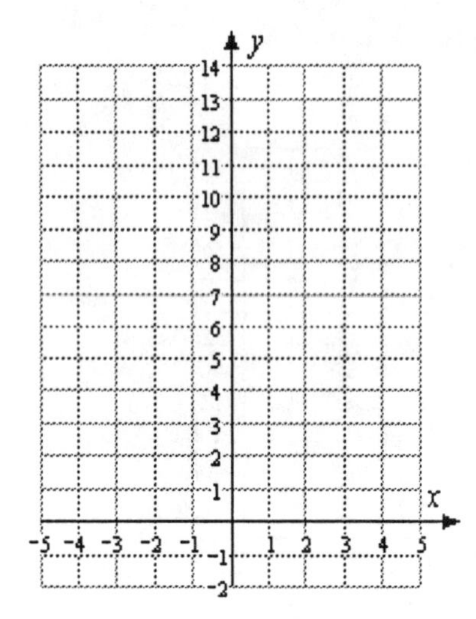

8. a. $y = (x-2)^2$
   b. $y = (x+3)^2$

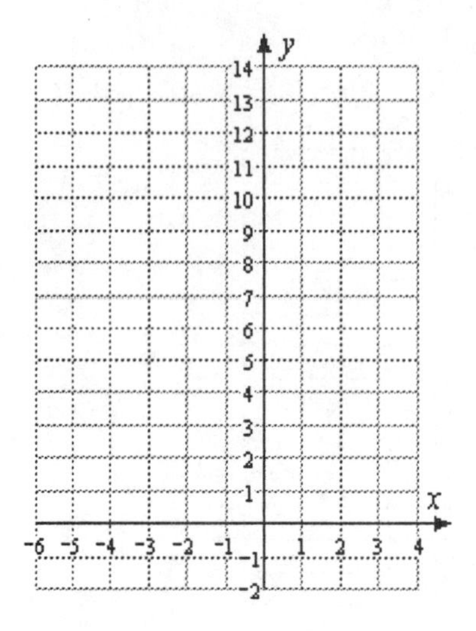

9. a. $y = -(x+1)^2$
   b. $y = -(x-4)^2$

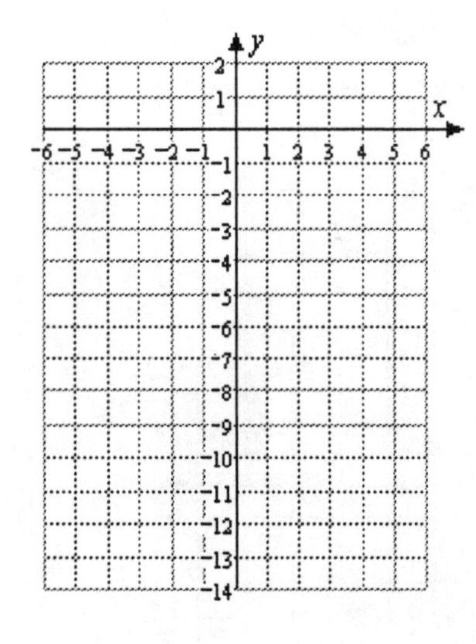

10. a. $y = -(x-3)^2$
    b. $y = -(x+4)^2$

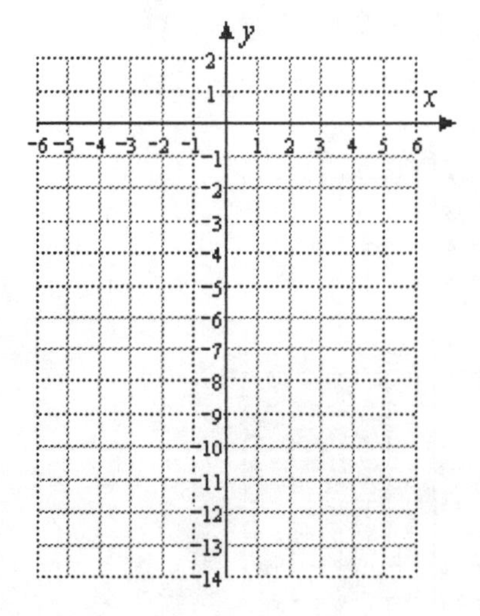

☐ *a. Graph each quadratic equation on the grid provided.*
b. What is the vertex of the parabola?
c. Write the equation in the form $y = ax^2 + bx + c$

13. $y = 2 + (x - 3)^2$

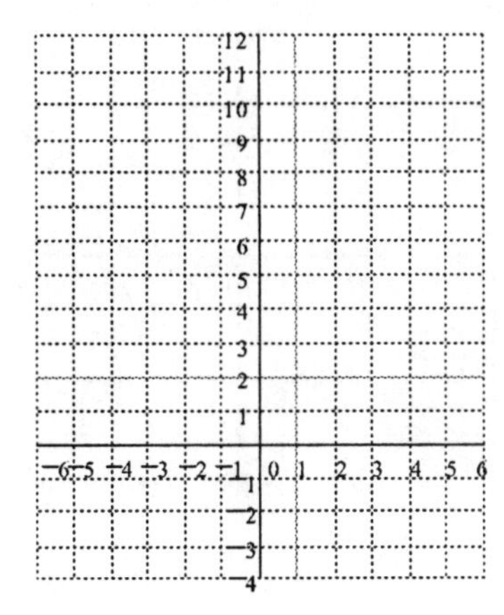

14. $y = -3 + (x + 4)^2$

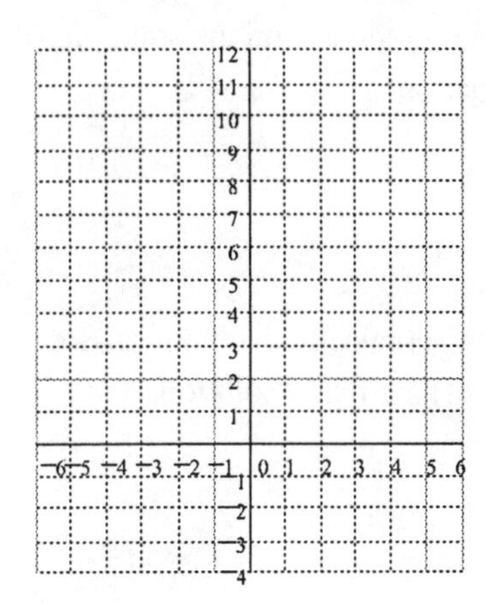

15. $y = 4 - (x + 1)^2$

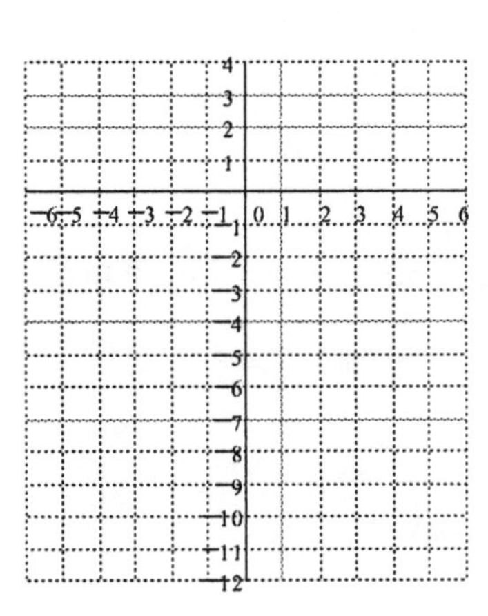

16. $y = -1 - (x - 2)^2$

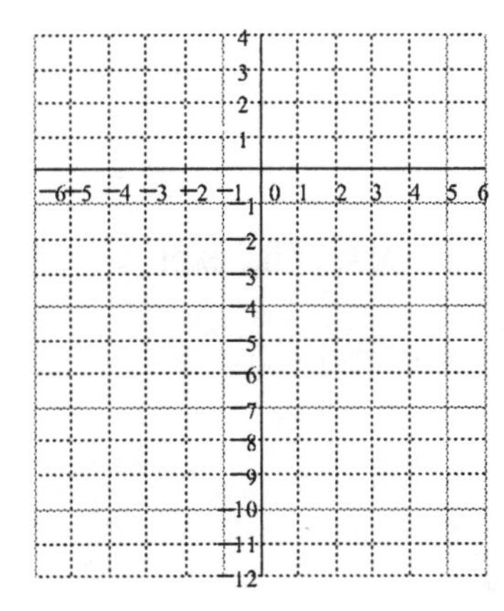

17. $y = x^2 + 2x - 8$.

x	-5	-4	-3	-2	-1	0	1	2	3
y									

a. Use your graph to find the solutions of the equation $x^2 + 2x - 8 = 0$.

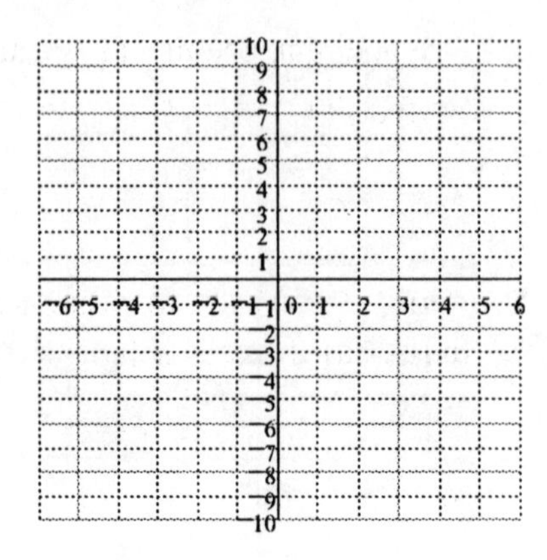

b. Verify algebraically that your answers to part (a) are really solutions.

c. Compute the product $(x+4)(x-2)$.

d. What is the value of the expression $(x+4)(x-2)$ when $x = -4$? When $x = 2$?

18. $y = x^2 - x - 6.$

x	-4	-3	-2	-1	0	1	2	3	4
y									

a. Use your graph to find the solutions of the equation $x^2 - x - 6 = 0$.

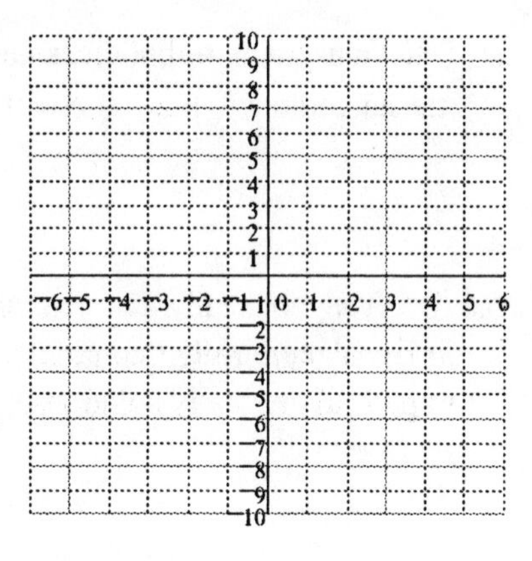

b. Verify algebraically that your answers to part (a) are really solutions.

c. Compute the product $(x-3)(x+2)$.

d. What is the value of the expression $(x-3)(x+2)$ when $x = 3$? When $x = -2$?

19. $y = -\ x^2\ +6x.$

x	-1	0	1	2	3	4	5	6	7
y									

a. Use your graph to find the solutions of the equation $-\ x^2\ +6\,x=0$.

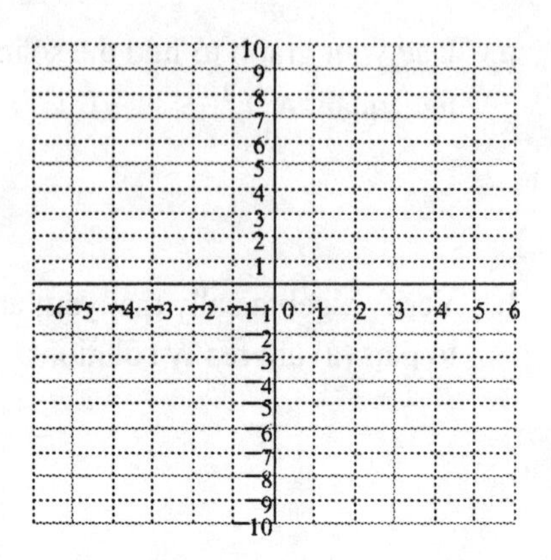

b. Verify algebraically that your answers to part (a) are really solutions.

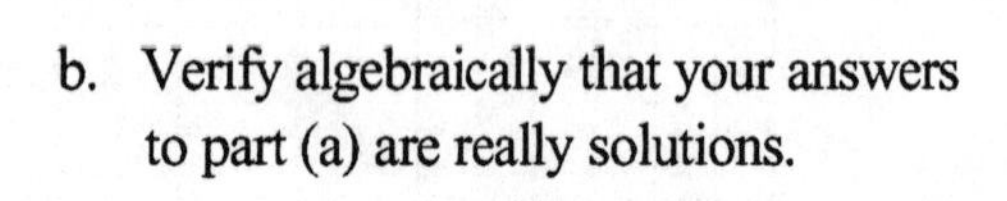

c. Compute the product $-x(x-6)$.

d. What is the value of the expression $-x(x-6)$ when $x=0$? When $x=6$?

20. $y = x^2 + 4x$.

x	-6	-5	-4	-3	-2	-1	0	1	2
y									

a. Use your graph to find the solutions of the equation $x^2 + 4x = 0$.

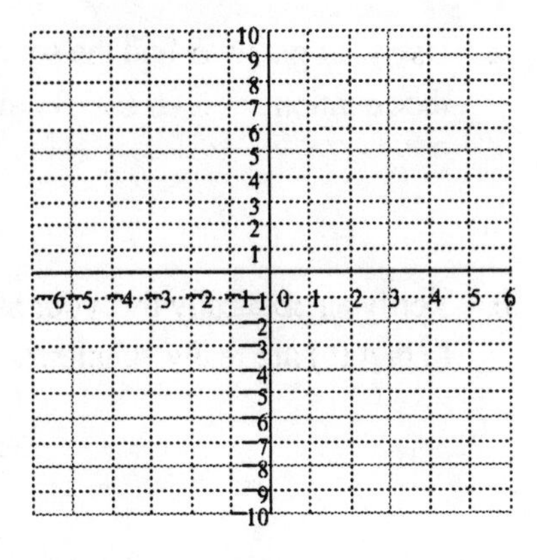

b. Verify algebraically that your answers to part (a) are really solutions.

c. Compute the product $x(x+4)$.

d. What is the value of the expression $x(x+4)$ when $x = 0$? When $x = -4$?

21. $y = x^2 - 6x + 9.$

x	-1	0	1	2	3	4	5	6	7
y									

a. Use your graph to find the solutions of the equation $x^2 - 6x + 9 = 0$.

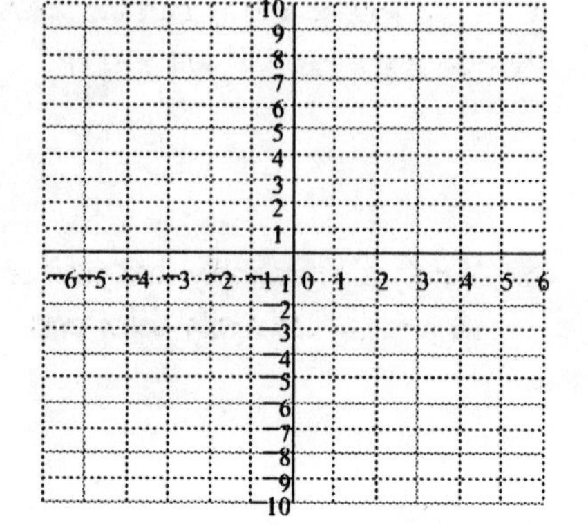

b. Verify algebraically that your answers to part (a) are really solutions.

c. Compute the product $(x-3)^2$.

d. What is the value of the expression $(x-3)^2$ when $x = 3$?

22. $y = 9 - x^2$.

x	-4	-3	-2	-1	0	1	2	3	4
y									

a. Use your graph to find the solutions of the equation $9 - x^2 = 0$.

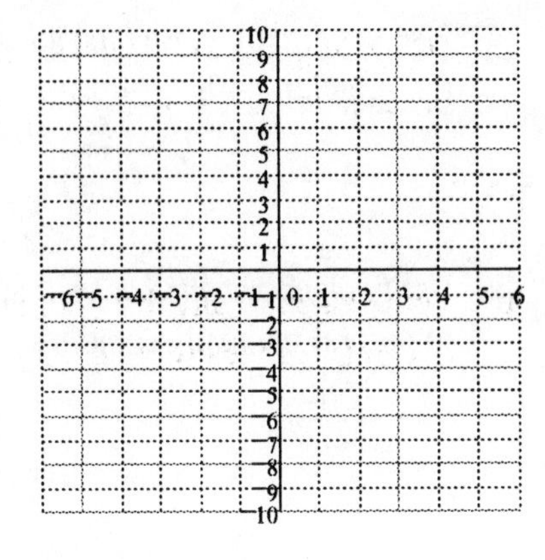

b. Verify algebraically that your answers to part (a) are really solutions.

c. Compute the product $(x+3)(x-3)$.

d. What is the value of the expression $(x+3)(x-3)$ when $x = 3$? When $x = -3$?

23. $y = -2x^2 + 12x - 10.$

x	-1	0	1	2	3	4	5	6	7
y									

a. Use your graph to find the solutions of the equation $-2x^2 + 12x - 10 = 0$.

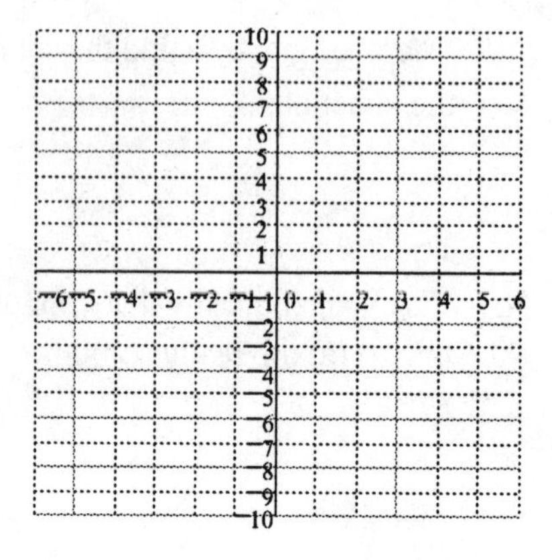

b. Verify algebraically that your answers to part (a) are really solutions.

c. Compute the product $-2(x-1)(x-5)$.

d. What is the value of the expression $-2(x-1)(x-5)$ when $x = 1$? When $x = 5$?

24. $y = \frac{1}{2}x^2 + x - 12.$

x	-7	-6	-5	-4	-3	-2	-1	0	1	2	3	4
y												

a. Use your graph to find the solutions of the equation $\frac{1}{2}x^2 + x - 12 = 0$.

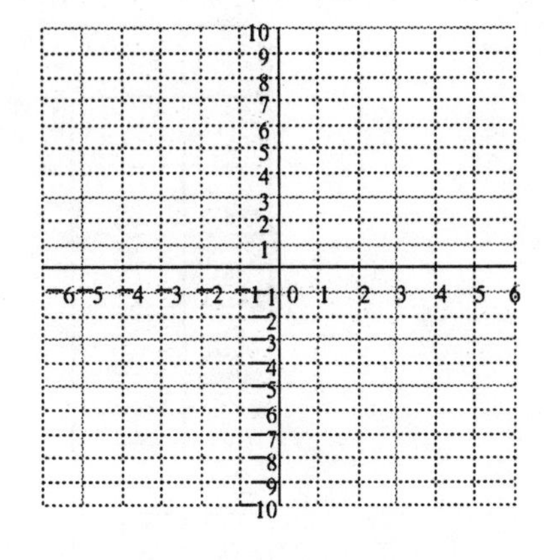

b. Verify algebraically that your answers to part (a) are really solutions.

c. Compute the product $\frac{1}{2}(x+6)(x-4)$.

d. What is the value of the expression $\frac{1}{2}(x+6)(x-4)$ when $x = -6$? When $x = 4$?

27. The bridge over the Rushing River at Marionville is 48 feet high. Francine stands on the bridge and tosses a rock into the air off the edge of the bridge. The height of the rockabove the water t seconds later is given in feet by

$$h = 48 + 32t - 16t^2.$$

a. Complete the table of values.

t	h
0	
0.5	
1	
1.5	
2	
2.5	
3	

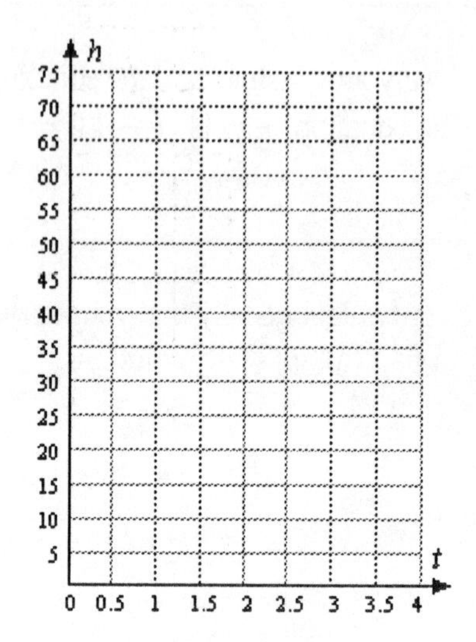

b. Sketch a graph of the equation on the grid.

c. Estimate the height of the rock after 1.75 seconds. Verify your answer algebraically.

d. When is the rock about 40 feet above the water?

e. Write an equation for the question in part (d).

f. How long is the rock more than 60 feet high?

g. After reaching its highest point, how long is the rock falling before it hits the water?

28. In a fireworks display, one of the rockets is shot from ground level with an initial speed of 96 feet per second. Its height t seconds later is given in feet by

$$h = 96t - 16t^2.$$

a. Complete the table of values.

t	h
0	
1	
2	
3	
4	
5	
6	

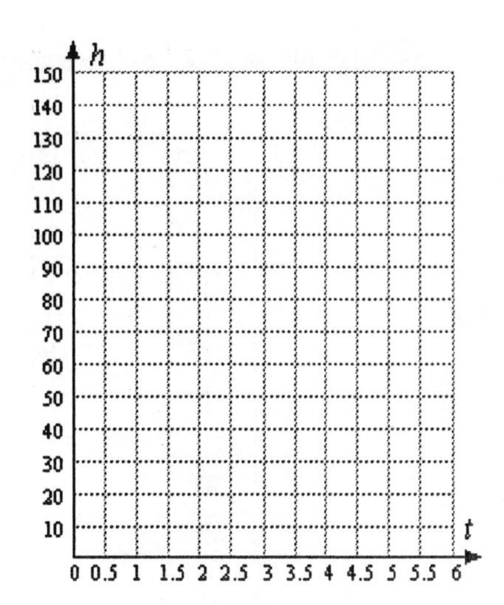

b. Sketch the graph of the equation on the grid.

c. Estimate the height of the rocket after 1.75 seconds. Verify your answer algebraically.

d. Find two times when the rocket is at a height of 140 feet.

e. Write an equation for the question in part (d).

f. How long does it take the rocket to reach its highest point?

g. How long does the rocket fall from its highest point back to earth?

29. Kitchenware Appliances finds that the cost of producing x toasters per week is given in dollars by

$$C = 0.1x^2 - 8x + 400.$$

a. Complete the table of values.
b. Sketch the graph of the equation on the grid.

x	C
10	
20	
30	
40	
50	
60	
70	
80	
90	
100	

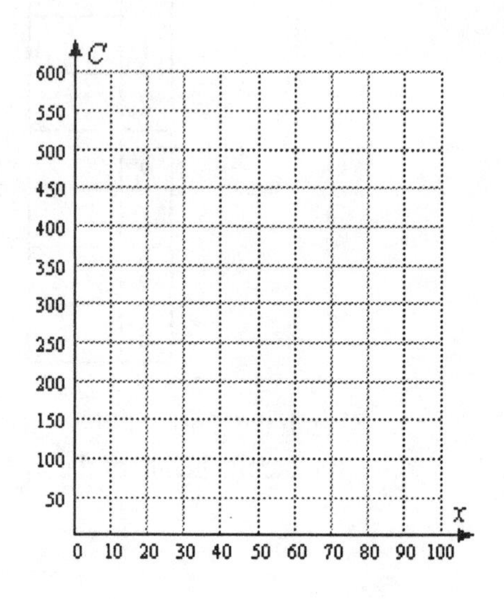

c. How much will it cost to produce 25 toasters per week?

d. How many toasters can be produced for \$330? (There are two answers.)

e. Write an equation for the question in part (d).

f. How many toasters should be produced if Kitchenware wants to minimize its costs?

g. How many toasters can be produced if costs must be kept under \$500?

30. Mitra plans to build a rectangular pen for her rabbits in the back yard. She has 48 feet of wire fence. If she builds a pen of width w feet, then the area of the pen is given in square feet by

$$A = 24w - w^2.$$

a. Complete the table of values.
b. Sketch a graph of the equation on the grid.

w	A
0	
4	
8	
12	
16	
20	
24	

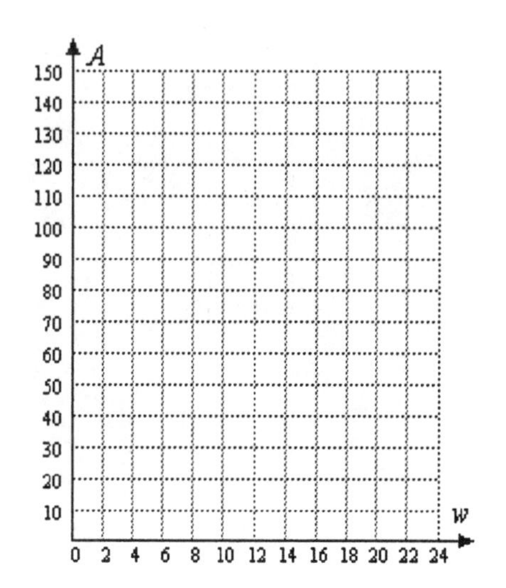

c. If the width of the pen is 10 feet, what is its area?

d. If the area of the pen is 120 square feet, what is its width? (There are two answers.)

e. Write an equation for the question in part (d).

f. What is the largest area that Mitra can enclose with her fence?

g. What are the dimensions of the largest rabbit pen Mitra can build?

Section 4.7 Solving Quadratic Equations

Exercise 1 Find the solutions of each quadratic equation.

a. $(x-2)(x+5)=0$ *b.* $y(y-4)=0$

Answers: a. *b.*

Exercise 2 Factor each expression.

a. $2x^2+8x$ *b.* $-3n^2+18n$

Answers: a. *b.*

Exercise 3 Solve each quadratic equation.

a. $2x^2+8x=0$ *b.* $-3n^2+18n=0$

Answers: a. *b.*

Exercise 4a. Find the x-intercepts of the graph of $y=2x^2+8x$.

b. Sketch the graph.

Answers a.

b.

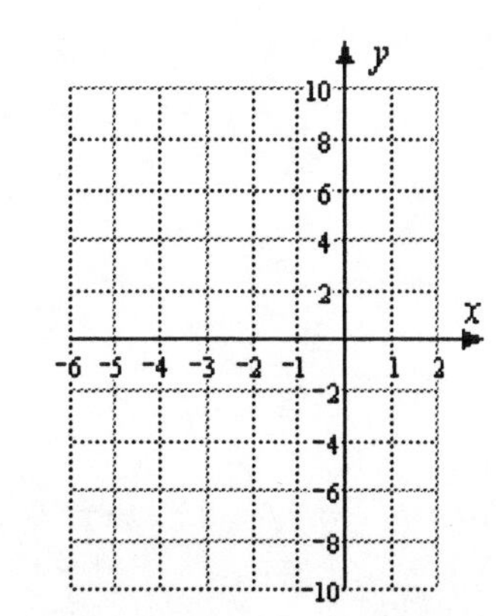

Homework 4.7

31. The revenue that you can earn by making and selling n jade bracelets is given in dollars by

$$R = -2n^2 + 80n.$$

a. Factor the expression for R.

b. Complete the table and sketch a graph.

n	R
0	
5	
10	
20	
30	
40	

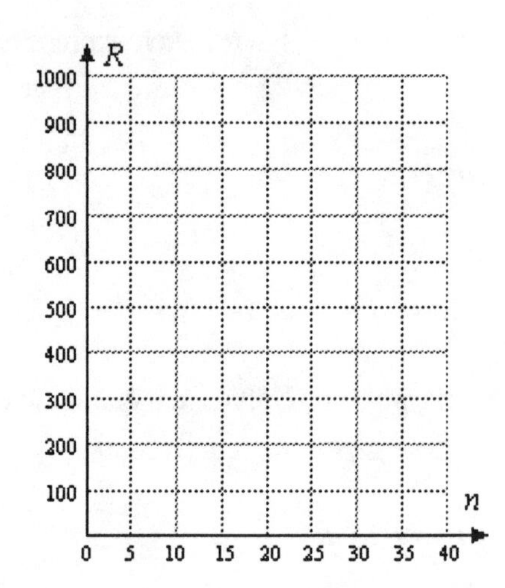

c. What is the maximum revenue that you can earn? How many bracelets should you make in order to earn that revenue?

32. The revenue earned (in dollars) from making and selling g glasses of lemonade is given by

$$R = -0.10g^2 + 4.80g.$$

a. Factor the expression for R.

b. Complete the table and sketch the graph.

g	R
0	
6	
12	
18	
24	
30	
36	
42	
48	

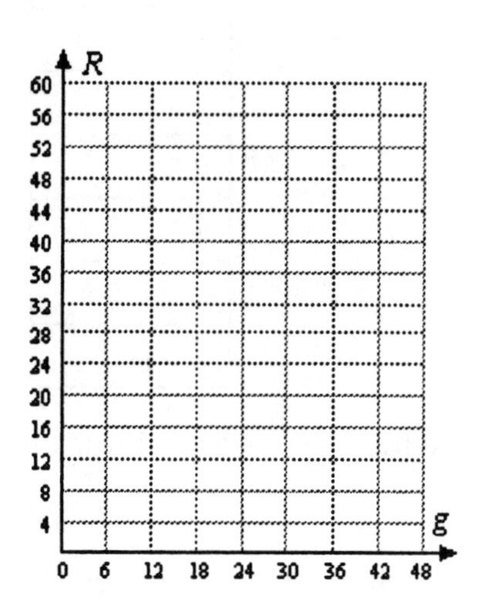

c. What is the maximum revenue that you can earn? How many glasses of lemonade should you make in order to earn that revenue?

33. The height in feet of a football t seconds after being kicked from the ground is given by

$$h = -16t^2 + 80t.$$

a. Factor the expression for h.

b. Complete the table and sketch the graph.

t	h
0	
1	
2	
2.5	
3	
4	
5	

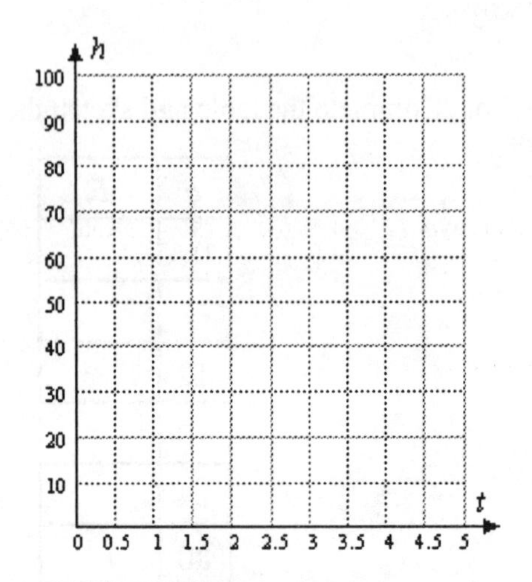

c. What is the maximum height of the football? When does it reach this height?

34. The height in centimeters of a golf ball t seconds after being hit is given by

$$h = -490t^2 + 1470t.$$

a. Factor the expression for h.

b. Complete the table and sketch the graph.

t	h
0	
0.5	
1	
1.5	
2	
2.5	
3	

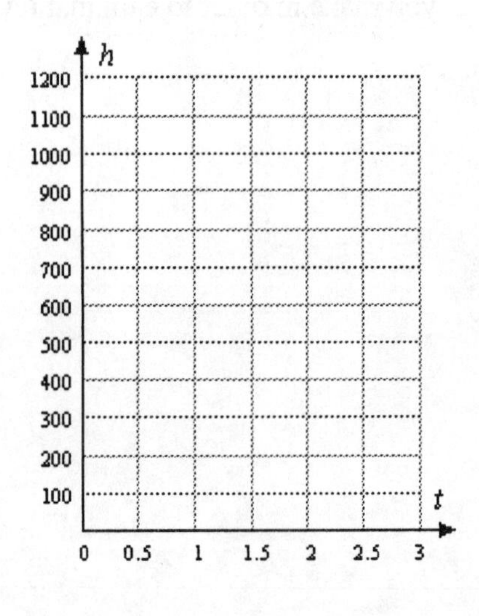

c. What is the maximum height of the golf ball? When does it reach this height?

Section 4.8 Factoring Quadratic Trinomials

Exercise 1 Factor each trinomial.

a. $x^2 + 8x + 15$ *b.* $y^2 + 14y + 49$

Answers: *a.* *b.*

Exercise 2 Factor each trinomial.

a. $m^2 - 10m + 24$ *b.* $m^2 - 11m + 24$

Answers *a.* *b.*

Exercise 3 Factor each trinomial.

a. $t^2 + 8t - 48$ *b.* $t^2 - 8t - 48$

Answers *a.* *b.*

Exercise 4 Solve each quadratic equation.

a. $a^2 - 13a + 30 = 0$ *b.* $u^2 - 6u = 16$

c. $3x^2 + 3 = 6x$ *d.* $9x^2 - 18x = 0$

Answers *a.* *b.*

c. *d.*

Exercise 5 Find the vertex of the parabola in Figure 4.34, whose equation is $y = -x^2 - 2x + 8$. (The x-intercepts are $(-4,0)$ and $(2,0)$.)

Answer:

Exercise 6a. Find the vertex and the y-intercept of the parabola in Example 5, $y = x^2 + 4x - 5$.

b. Sketch a graph of the equation on the grid.

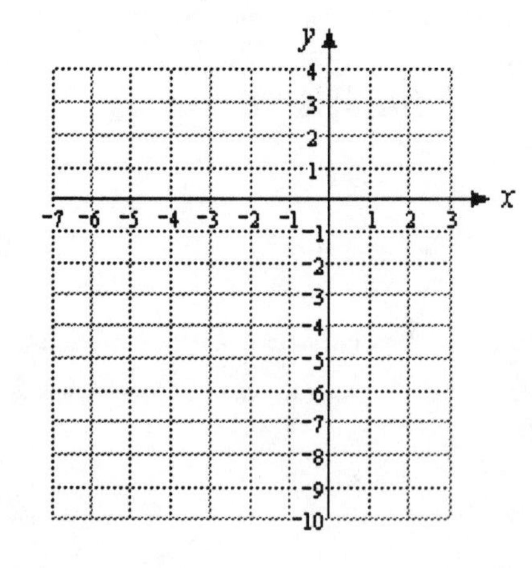

Answer:

Section 4.9 More About Factoring

Exercise 1 Compute each product by using the area of a rectangle, and verify that the products of the diagonal entries are equal.

a. $(2x-5)(3x-4)$ ***b.*** $(4t+15)(t-6)$

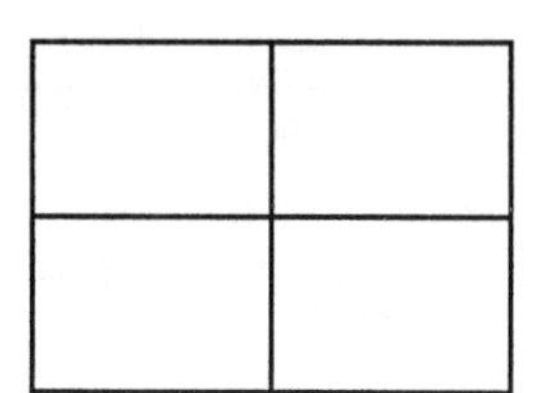

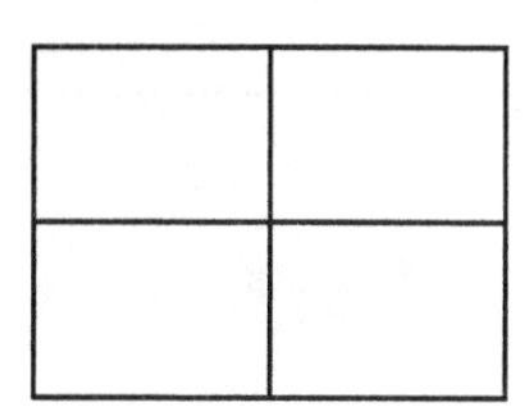

Answers: ***a.*** ***b.***

Exercise 2 Factor $4x^2 + 4x - 3$.

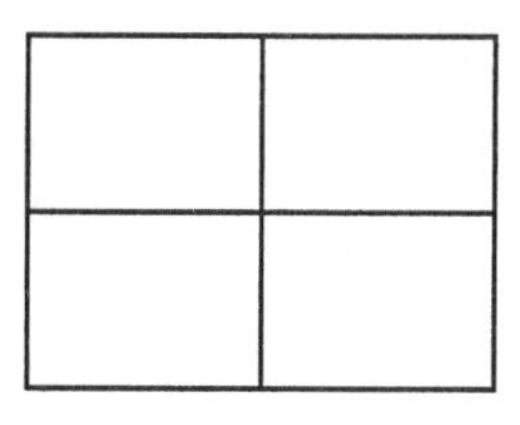

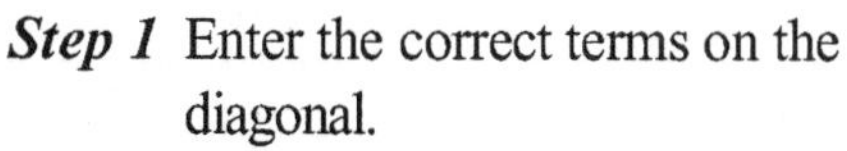

Step 1 Enter the correct terms on the diagonal.

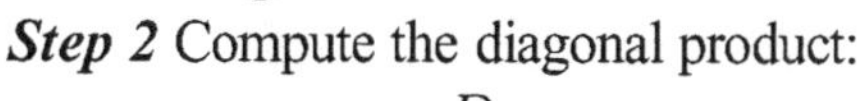

Step 2 Compute the diagonal product:

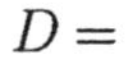

$D =$

Step 3 List all possible factors of D, and compute the sum of each pair of factors. Note that the factors must have opposite signs. (Complete each pair of factors below.)

Factors of $D =$		***Sum of Factors***
$-x$	____	
$-2x$	____	
$-3x$	____	

The correct factors are:

Step 4 Enter the correct factors into the rectangle.

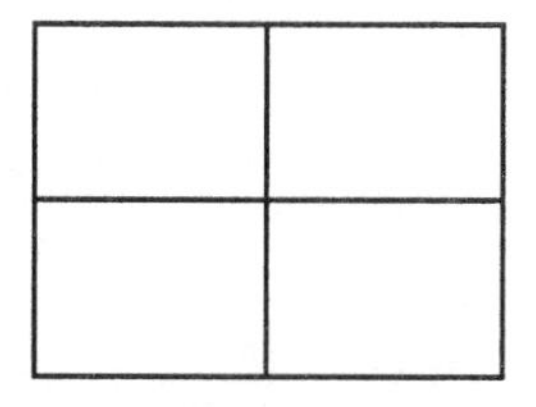

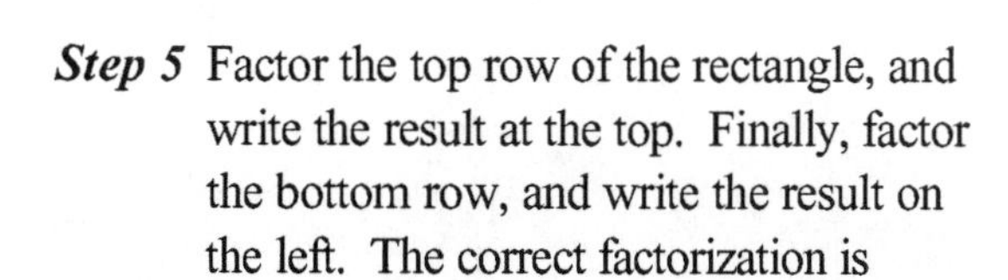

Step 5 Factor the top row of the rectangle, and write the result at the top. Finally, factor the bottom row, and write the result on the left. The correct factorization is

$$4x^2 + 4x - 3 =$$

Exercise 3 Use a rectangle to factor each quadratic trinomial.

a. $3z^2 + 10z + 3$ ***b.*** $4p^2 - 11p + 6$

Answers

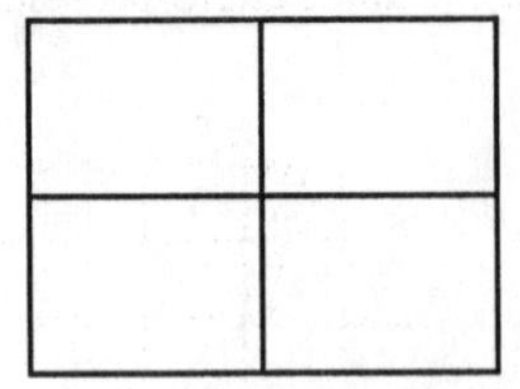

a. ***b.***

Exercise 4 Solve $2x^2 = 7x + 15$.

Answer First, write the equation in standard form: ____________________

Factor the left side. (Use the rectangle provided.) ____________________

Factors of D ***Sum of Factors***

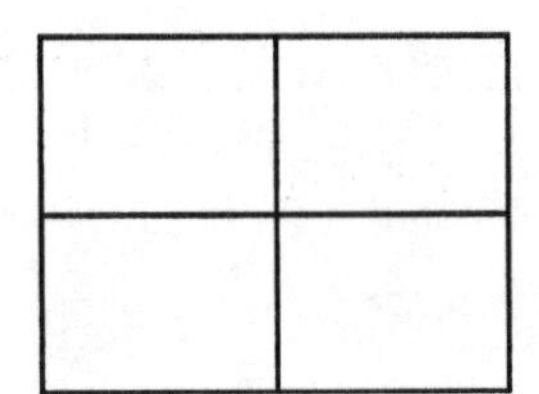

Set each factor equal to zero. __________________________

Solve each equation.

Section 4.10 The Quadratic Formula

Exercise 1a. Here are equations for three parabolas. Factor each formula if possible.

(I) $y = x^2 - 4x + 3$

(II) $y = x^2 - 4x + 4$

(III) $y = x^2 - 4x + 5$

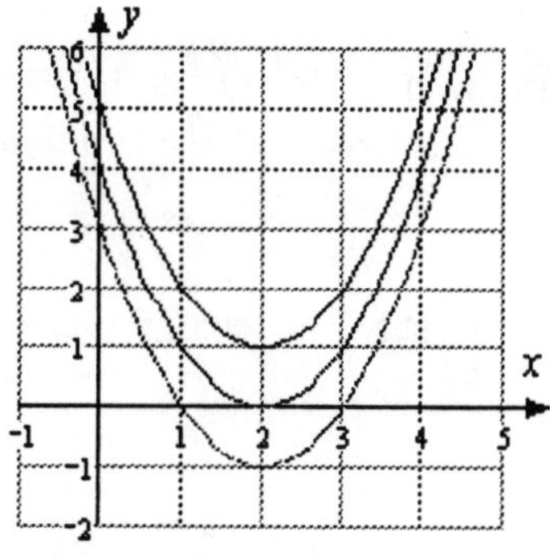

Figure 4.36

b. Match each equation from part (a) with one of the graphs shown. Explain your reasoning.

Exercise 2a. Use the quadratic formula to solve $2x^2 = 7 - 4x$.

b. Find decimal approximations to two decimal places for the solutions.

Answers: ***a.*** First write the equation in standard form: ____________________

Identify the coefficients: $a =$ ______, $b =$ ______, $c =$ ______.

Substitute the values of a, b, and c into the quadratic formula:

$$x = \frac{-b \pm \sqrt{b^2 - 4ac}}{2a} =$$

Simplify the expression. Start with the expression under the radical.

b. Use a calculator to approximate each solution.

Exercise 3a. Solve $x = \frac{2}{3} - \frac{x^2}{6}$.

b. Find decimal approximations to two decimal places for the solutions.

Answers: a. Multiply each term of the equation by 6, then write it in standard form.

Identify the coefficients and apply the quadratic formula.

$$x = \frac{-b \pm \sqrt{b^2 - 4ac}}{2a} =$$

b. Use a calculator to approximate each solution.

Exercise 4 Give as much information as you can about the x-intercepts of each parabola.

a. $y = x^2 + 4$

b. $y = 3x^2 + 4x - 2$

c. $y = -x^2 + 2x - 1$

d. $y = 2x^2 + 4x + 3$

Answers: a.

b.

c.

d.

Chapter 4 Summary and Review

39. If you are standing at an altitude of h meters, then the distance d that you can see to the horizon is given in miles by

$$d = \sqrt{12h}.$$

a. Complete the table, and graph the equation on the grid.

h	d
0	
1000	
2000	
3000	
4000	
5000	

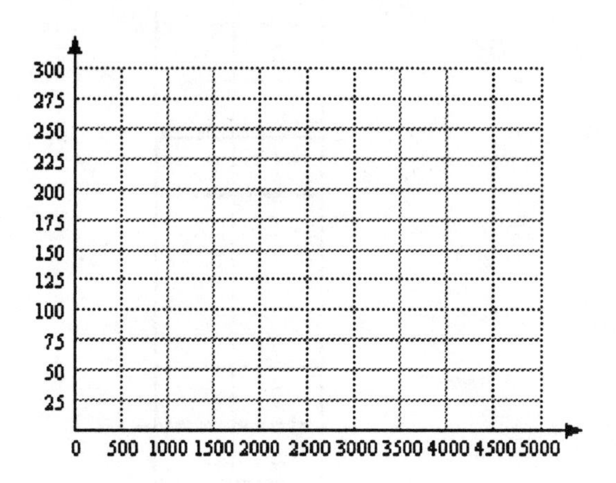

b. Use your graph to estimate the distance you can see from the top of Mt. Whitney, at an altitude of 4,149 meters.

c. Use the equation to find the distance in part (b) algebraically.

d. If you want to see for 100 miles, what altitude do you need?

59. If you drop a stone from a bridge 100 feet above the water, the height h of the stone t seconds after you drop it is given in feet by

$$h = 100 - 16t^2.$$

a. Complete the table.

t	h
0	
0.5	
1	
1.25	
1.5	
1.75	
2	
2.25	
2.5	

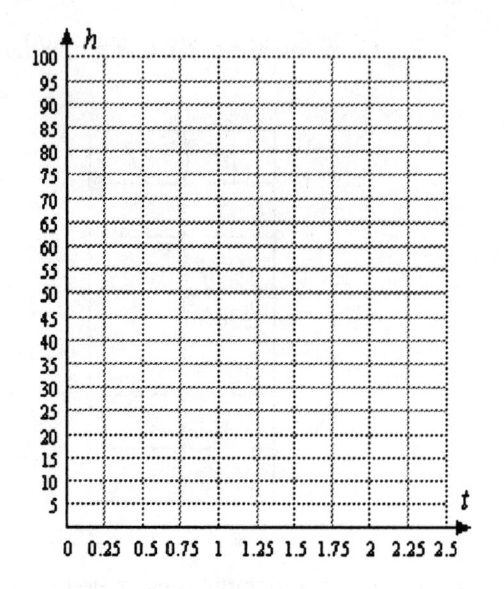

b. Sketch the graph of the equation on the grid.

c. What is the height of the stone after 2 seconds? Verify your answer algebraically.

d. When is the height of the stone 75 feet? Write an algebraic equation you can solve to verify your answer.

60. Sportsworld sells $180-3p$ pairs of their name-brand running shoes per week when they charge p dollars per pair.

a. Write an equation for Sportsworld's revenue, R, in terms of p.

b. Fill in the table, and graph your equation on the grid.

p	R
0	
10	
20	
30	
40	
50	
60	

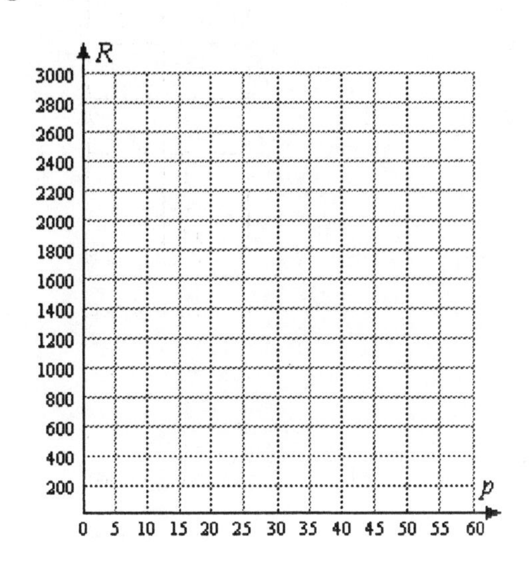

c. At what price(s) will Sportsworld's revenue be zero?

d. At what price will Sportsworld's revenue be maximum? What is their maximum revenue?

Chapter 5 Polynomials and Rational Expressions

Section 5.1 Polynomials

Exercise 1 Which of the following expressions are polynomials?

a. $t^2 - 7t + 3$ *b.* $2 + \frac{6}{x}$

c. $-a^3b^2c^4$ *d.* $3w^2 - w + 5\sqrt{w}$

Answers:

Exercise 2 Give the degree of each polynomial.

a. $5x^9 - 1$ *b.* $3t^2 - t - 6t^4 + 8$

Answers: *a.* *b.*

Exercise 3 Write the polynomial $3x - 8 - 6x^3 - x^6$ in descending powers of x.

Answer:

Exercise 4 Evaluate $2x^3y^2 - 3xy^3 - 3$ for $x = -2,\ y = -3$.

Answer:

Exercise 5 Simplify each expression by combining any like terms.

a. $x^3 + 2x^2 - 2x^3 - (-4x^2) + 4x$ *b.* $2a^2b - 5ba + 4ba^2 - (-3ab)$

Answers: *a.* *b.*

Exercise 6 Add the polynomials.

a. $(4x^2 - 3x - 1) + (5x^2 + x - 1)$

b. $(2m^4 - 4m^3 + 3m - 1) + (m^3 + 4m^2 - 6m + 2)$

Answers:

a.

b.

Exercise 7 Subtract the polynomials.

a. $(y^2 - 3y + 5) - (y^2 + 4y - 3)$

b. $(2m^4 - 4m^3 + 3m - 1) - (m^3 + 4m^2 - 6m + 2)$

Answers:

a.

b.

Exercise 8 Use a vertical format to add or subtract the polynomials.

a. Add $5y^2 - 3y + 1$ to $-3y^2 + 6y - 7$.

b. Subtract $2x^2 + 5 - 2x$ from $7 - 3x - 4x^2$.

Answers:

a.

b.

Homework 5.1

69. Let S_n stand for the sum of the first n integers. For example,

$$S_1 = 1,$$
$$S_2 = 1 + 2 = 3,$$
$$S_3 = 1 + 2 + 3 = 6,$$

and so on.

a. Fill in the table showing the first 10 values of S_n.

b. Evaluate the polynomial $\frac{1}{2}n^2 + \frac{1}{2}n$ for integer values of n from 1 to 10, and fill in the table.

c. Compare your answers to parts (a) and (b).

n	S_n	$\frac{1}{2}n^2 + \frac{1}{2}n$
1		
2		
3		
4		
5		
6		
7		
8		
9		
10		

70. Let T_m stand for the sum of the squares of the first m integers. For example,

$$T_1 = 1^2 = 1,$$
$$T_2 = 1^2 + 2^2 = 5,$$
$$T_3 = 1^2 + 2^2 + 3^2 = 14,$$

and so on.

a. Fill in the table showing the first 10 values of T_m.

b. Evaluate the polynomial $\frac{1}{3}m^3 + \frac{1}{2}m^2 + \frac{1}{6}m$ for integer values of m from 1 to 10, and fill in the table.

c. Compare your answers to parts (a) and (b).

m	T_m	$\frac{1}{3}m^3 + \frac{1}{2}m^2 + \frac{1}{6}m$
1		
2		
3		
4		
5		
6		
7		
8		
9		
10		

Section 5.2 Products of Polynomials

Exercise 1 Use the first law of exponents to find each product.

a. $k^2 \cdot k^8$ *b.* $y^3(y^3)$

Answers *a.* *b.*

Exercise 2 Explain why each product below is *incorrect*. Then give the correct product.

a. $3^4 \cdot 3^3 \rightarrow 9^7$ *b.* $t^3 \cdot t^5 \rightarrow t^{15}$ ← **Incorrect!**

Answers *a.* *b.*

Exercise 3 Multiply $-3a^4b(-4a^3b)$.

Answer:

Exercise 4 Simplify $2a^2(3 - a + 4a^2) - 3a(5a - a^2)$ by following the steps below:

Step 1 Apply the distributive law to remove parentheses.

Step 2 Use the First Law of Exponents to simplify each term.

Step 3 Combine like terms.

Exercise 5 Multiply $(3x+2)(3x^2+4x-2)$

Answer:

Exercise 6 Multiply $s^2t^2(2s+1)(3s-1)$.

Answer:

Exercise 7 Multiply $(x+2)(3x-2)(2x-1)$.

Answer:

Section 5.3 Factoring Polynomials

Exercise 1 Use the Second Law of Exponents to find each quotient.

a. $\frac{x^6}{x^2}$ ***b.*** $\frac{b^2}{b^3}$

Answers: ***a.*** ***b.***

Exercise 2 Divide $\frac{8x^2y}{12x^5y^3}$.

Answer:

Exercise 3 Find the greatest common factor for $15x^2y^2 - 12xy + 6xy^3$

Answer:

Exercise 4 Follow the steps given to factor $15x^2y^2 - 12xy + 6xy^3$.

Answer:

Step 1 Find the GCF:

Step 2 Write the GCF outside a set of parentheses:

Step 3 Divide each term by the GCF:

Step 4 Write the quotients inside the parentheses.

Exercise 5 Factor $9(x^2+5)-x(x^2+5)$.

Answer:

Exercise 6 Factor completely $4a^6-10a^5+6a^4$.

Answer:

Exercise 7 Factor completely.

a. $2a^3b-24a^2b^2-90ab^3$ ***b.*** $3x^2-8xy+4y^2$

Answers: a. ***b.***

Section 5.4 Special Products and Factors

Exercise 1 Simplify each square.

a. $(6t^4)^2$ ***b.*** $(12st^8)^2$

Answers: ***a.*** ***b.***

Exercise 2 Find a monomial whose square is $64b^6$.

Answer:

Exercise 3 Use formula 2 to expand $(4-3t)^2$.

Answer:

Exercise 4 Multiply $(5x^4+4)(5x^4-4)$.

Answer:

Exercise 5 Use one of the three formulas to factor each polynomial.

a. $25y^2 - w^2$ ***b.*** $m^6 - 18m^3 + 81$

Answers

Step 1 Choose the appropriate formula.

Step 2 Identify a and b.

Step 3 Verify that the middle term (if any) equals $2ab$.

Step 4 Replace a and b in the formula with your answers to Step 2.

Exercise 6 Factor completely $x^6 - 16x^2$.

Answer:

Section 5.5 Inverse Variation

Exercise 1 Use the equation $p=1.5g$ to complete the table. For each row, check that the new value of p is double the old value.

Answers:

g	p	**Double** g	**Compute new** p
2			
3			
5			

Exercise 2 Arlen would like to build a rectangular rabbit pen with area 60 square feet. If the length and width of the pen are l and w, then we know that

$$lw=60.$$

a. Complete the table.

b. Use the table to answer the following questions:
What happens to the value of l if you double the value of w?

What happens to the value of l if you triple the value of w?

w	l
2	
3	
4	
5	
6	
10	
12	
15	

Activity

1. In Exercise 2 you made a table of possible lengths and widths for a rectangle of area 60 square feet.
   b. Write an equation for the length, l, of the rectangle in terms of its width, w.

      $l =$

   c. Use the table to graph your equation on the grid in Figure 5.4.
   d. What will happen to the value of l if you continue to increase the value of w?

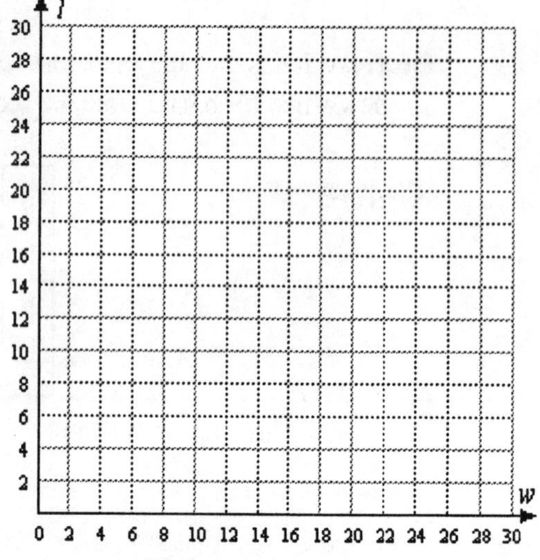

Figure 5.4

2. Matt wants to travel 360 miles to visit a friend over spring break. He is deciding whether to ride his bike or drive. If he travels at an average speed of r miles per hour, then the trip will take t hours.
   a. Fill in the table.

r	t
10	
20	
30	
40	
60	
80	
90	

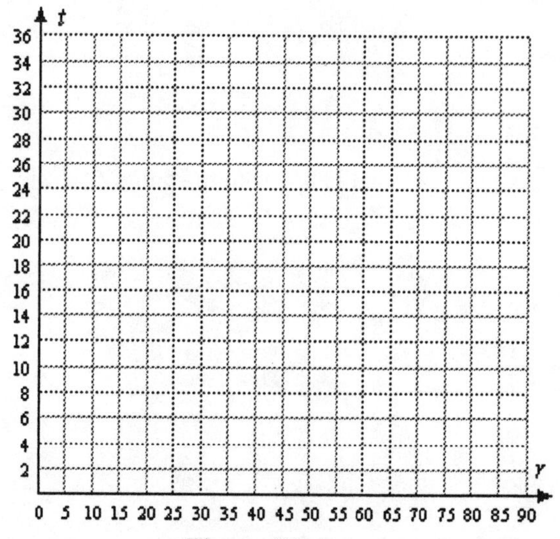

Figure 5.5

   b. Write an equation for t in terms of r.

      $t =$

   c. Use the table to graph your equation on the grid in Figure 5.5.
   d. If Matt doubles his speed, what happens to his travel time? Show this on your graph, starting with $r = 20$.

Exercise 3 y varies inversely with x, and $y = 12$ when $x = 4$.

a. What is the constant of variation?

b. What is the value of y when $x = 3$?

Answers ***a.*** ***b.***

Exercise 4 Look at the tables of values you completed in the activity. How can you recognize inverse variation from a table of values?

Answer:

Exercise 5 Which of the following tables represent inverse variation?

a.

h	1.6	2	3.2	8
w	6	4.8	3	1.2

b.

V	320	270	230	160
R	80	130	170	240

Homework 5.5

1. To celebrate their graduation, Janel and her friends want to charter a boat for a dinner cruise on the river. It costs $1200 to rent the boat for the evening (not including dinner), and they will split the cost equally.

   a. Fill in the table showing each person's share of the rental fee, s, if n people go on the cruise.

n	s
5	
6	
10	
12	
20	
25	
30	
40	

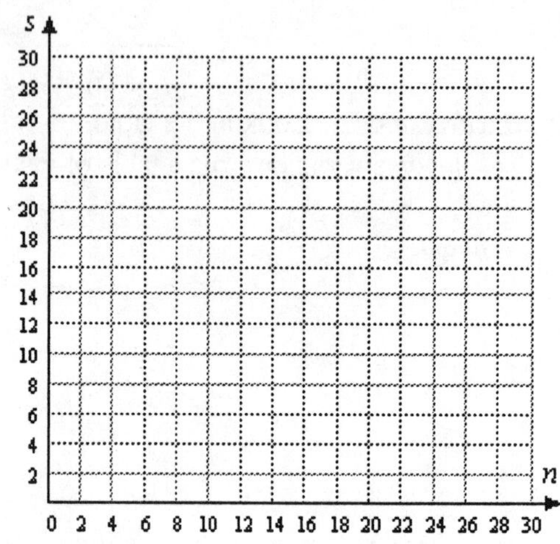

   b. Write an equation for s in terms of n.

$$s =$$

   c. Graph your equation on the grid.

   d. At the last minute, one-third of the people signed up for the cruise back out. By what factor does the price per person increase?

2. Every Friday, Takuya takes his company's weekly employee bulletin to the copy center and asks for 1000 copies. The time it takes to make the copies depends on the speed of the copier used.

   a. Fill in the table showing the time, t, to make the copies (in minutes) on a copier that produces c copies per minute.

c	t
5	
10	
20	
25	
50	
100	
125	

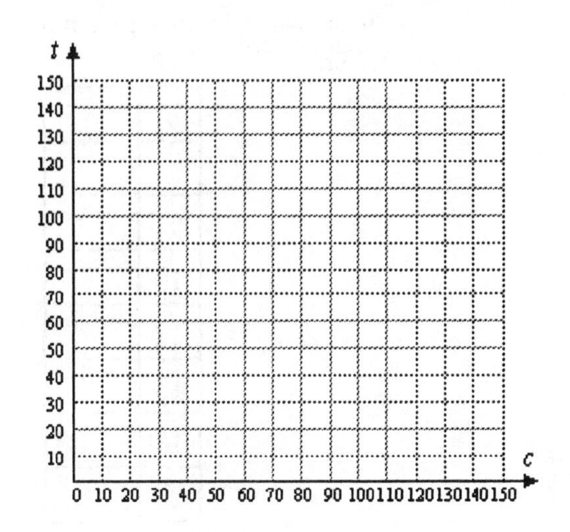

   b. Write an equation for t in terms of c.

$$t =$$

   c. Graph your equation on the grid.

   d. If the speed of the copier can be increased by 25%, by what factor will the time required decrease?

3. The manager at Cut 'n' Style finds that the number of customers per week varies inversely with the price he charges for a haircut. During a week when he charged $10, he had 36 customers.
   a. Find the constant of variation and write an equation for the number of customers, c, expected when the price of a haircut is p dollars.

      $c =$

   b. Fill in the table.

p	c
6	
9	
10	
12	
15	
18	
20	

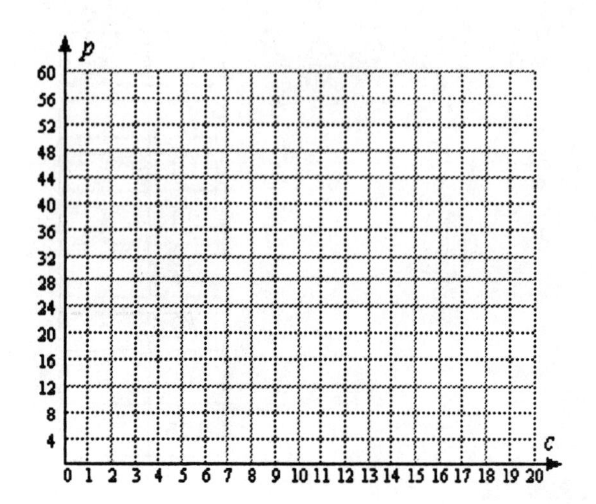

   c. Graph your equation on the grid in.
   d. If the manager increases the price of a haircut by 50%, by what factor will the number of customers decrease?

4. When two people balance on a see-saw, the distance each sits from the fulcrum varies inversely with his of her weight. Shannon weighs 120 pounds and is sitting 3 feet from the fulcrum. Each child in her preschool class gets a turn on the see-saw.
   a. If a child weighs w pounds, write an equation for the distance, d, the child should sit from the fulcrum.

$$d =$$

   b. Fill in the table.

w	d
30	
36	
40	
45	
60	

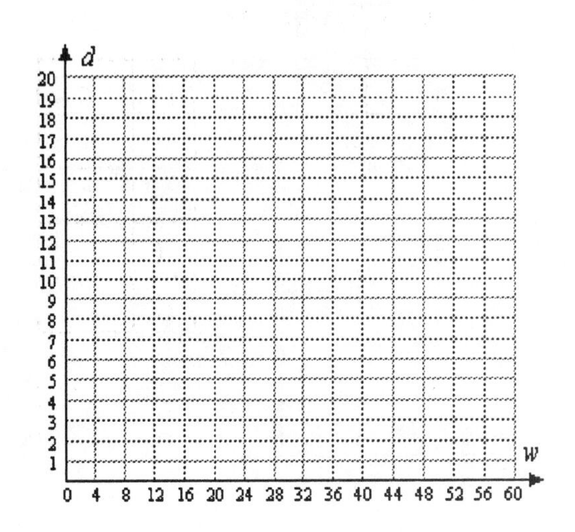

   c. Graph your equation on the grid.

   d. Shawn weighs $\frac{4}{3}$ of Shannon's weight. Where should he sit in order to balance the children at the same distances they are used to?

☐ *Each table represents inverse variation. Find the constant of variation, write an equation relating the variables, then complete the table.*

5.

w	t
4.5	8
7.2	
	4.8

6.

z	h
32	21
42	
	14

7.

R	C
20	200
	250
125	

8.

m	T
0.2	50
0.8	
	6.25

☐ *In Exercises 17-20, you will graph equations of the form* $y = \frac{k}{x}$.

a. *First consider large positive values of* x. *What happens to* y *as* x *gets large positive?*

b. *Next evaluate* y *for small positive values of* x. *What happens to* y *as* x *gets small positive?*

c. *Now look at negative values of* x. *What happens to* y *as* x *gets "large" negative?*

d. *Finally, consider negative* x *close to zero. What happens to* y *as* x *approaches* 0 *from the negative side?*

(Note that the expression $\frac{k}{x}$ *is undefined when* $x = 0$.*)*

17. Plot the points to graph $y = \frac{1}{x}$.

a.

x	2	4	5	8
y				

b.

x	1	$\frac{1}{2}$	$\frac{1}{4}$	$\frac{1}{8}$
y				

c.

x	-2	-3	-6	-8
y				

d.

x	-1	$-\frac{1}{2}$	$-\frac{1}{3}$	$-\frac{1}{5}$
y				

18. $y = \frac{-1}{x}$

a.

x	2	4	5	8
y				

b.

x	1	$\frac{1}{2}$	$\frac{1}{4}$	$\frac{1}{8}$
y				

c.

x	-2	-3	-6	-8
y				

d.

x	-1	$-\frac{1}{2}$	$-\frac{1}{3}$	$-\frac{1}{5}$
y				

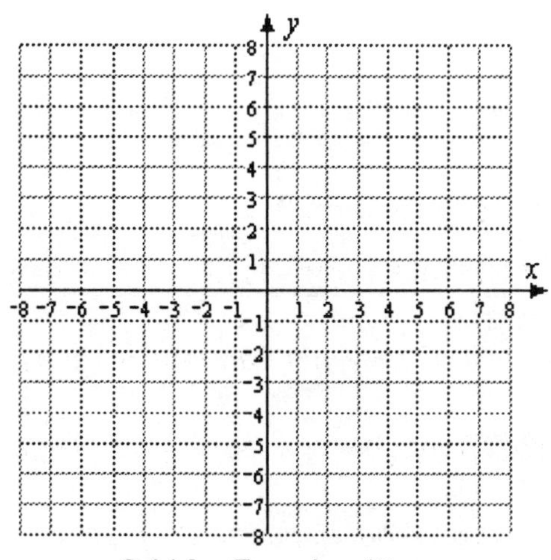

Grid for Exercise 17

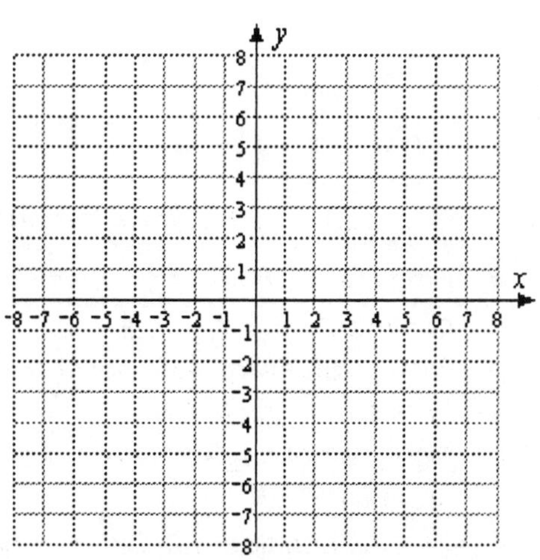

Grid for Exercise 18

19. $y = \frac{-3}{x}$

a.

x	2	4	5	8
y				

b.

x	1	$\frac{1}{2}$	$\frac{1}{4}$	$\frac{1}{8}$
y				

c.

x	-2	-3	-6	-8
y				

d.

x	-1	$-\frac{1}{2}$	$-\frac{1}{3}$	$-\frac{1}{5}$
y				

20. $y = \frac{2}{x}$

a.

x	2	4	5	8
y				

b.

x	1	$\frac{1}{2}$	$\frac{1}{4}$	$\frac{1}{8}$
y				

c.

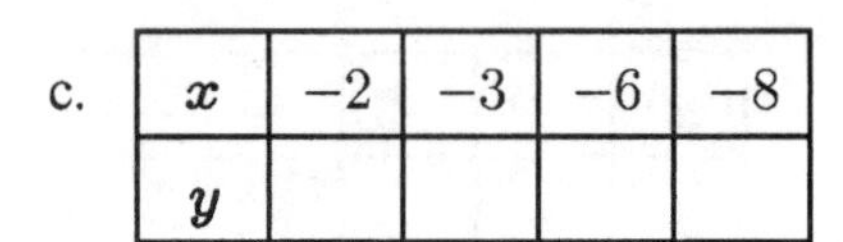

x	-2	-3	-6	-8
y				

d.

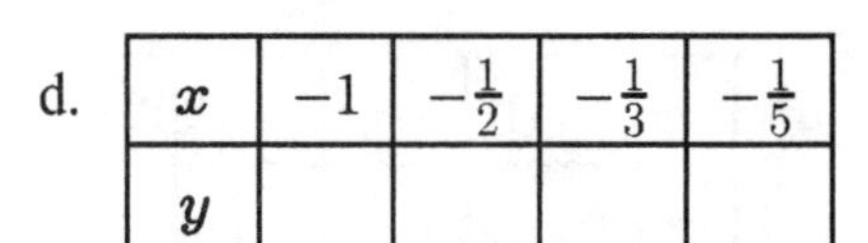

x	-1	$-\frac{1}{2}$	$-\frac{1}{3}$	$-\frac{1}{5}$
y				

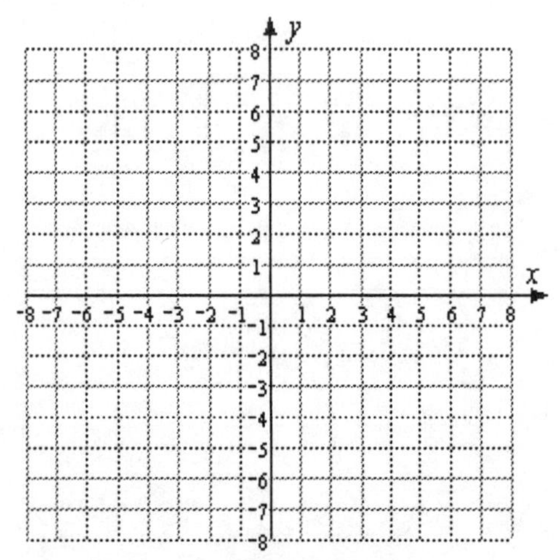

Grid for Exercise 19

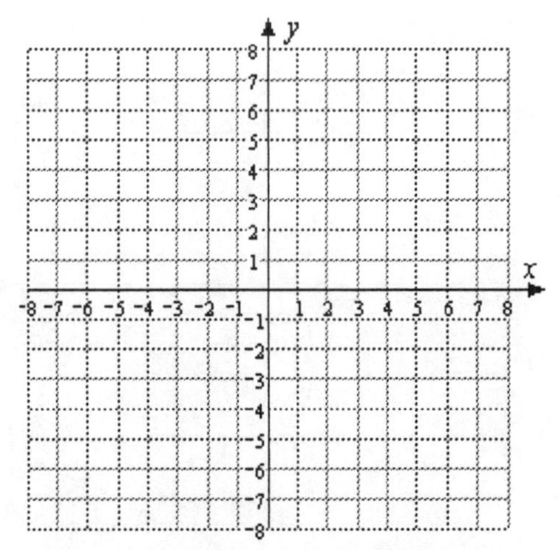

Grid for Exercise 20

Section 5.6 Algebraic Fractions

Exercise 1a. Evaluate $\dfrac{z-1}{2z+3}$ for $z=-3$.

b. For what value of z is the fraction undefined?

Answers *a.*

b.

x	A
50	150
100	100
200	75
400	62.50
500	60
1000	55
1250	54
2000	52.50

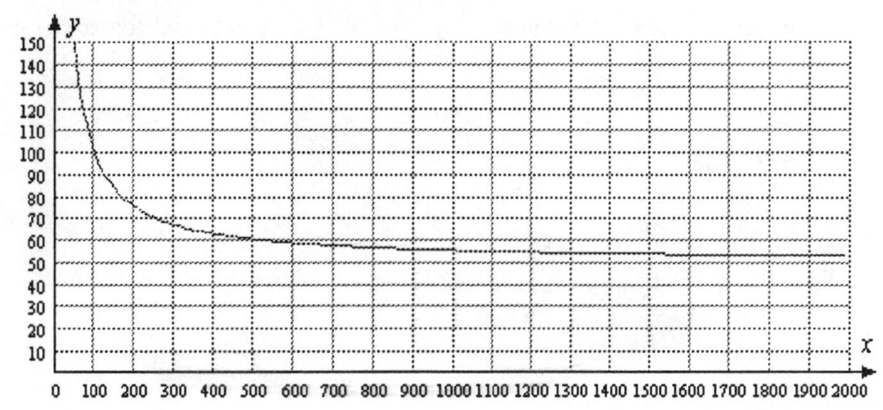

Figure 5.7

Exercise 2 Use the graph of average cost and the table to help you answer the questions.

a. Approximately how many filters should Envirogreen produce for the average cost to be \$75?

b. What happens to the average cost per filter as x increases?

c. As x increases, the average cost appears to be approaching a limiting value. What is that value?

d. A market analysis concludes that Envirogreen can sell 1250 filters this year. How much should they charge for one filter if they would like to make a total profit of \$100,000?

Answers: *a.* *b.*

c. *d.*

Exercise 3 Use your calculator to determine which calculation in each pair is correct. (In part b, choose a value for the variable and evaluate.)

a. $\dfrac{12}{8} = \dfrac{4 \cdot 3}{4 \cdot 2} \to \dfrac{3}{2}$ $\qquad$ $\dfrac{7}{6} = \dfrac{4+3}{4+2} \to \dfrac{3}{2}$

b. $\dfrac{5x}{8x} \to \dfrac{5}{8}$ $\qquad$ $\dfrac{x+5}{x+8} \to \dfrac{5}{8}$

Exercise 4 If you evaluate $\dfrac{3x+12}{6x+24}$ for $x = 2$ and for $x = -5$, what answer do you expect to get?

Answer:

Exercise 5 Reduce $\dfrac{x^2 - x - 6}{x^2 - 9}$.

Answer:

Exercise 6 Explain why each calculation is *incorrect*.

a. $\dfrac{x-6}{x-9} \to \dfrac{x-2}{x-3}$ $\qquad$ ***b.*** $\dfrac{2x+5}{2} \to x+5$

Answers: a.

b.

Exercise 7 Find the opposite of each binomial.

a. $2a-3b$ *b.* $2a+3b$ *c.* $-x-1$ *d.* $-x+1$

Answers *a.* *b.* *c.* *d.*

Exercise 8 Reduce if possible.

a. $\dfrac{-x+1}{1-x}$ *b.* $\dfrac{1+x}{1-x}$ *c.* $\dfrac{2a-3b}{3b-2a}$ *d.* $\dfrac{2a-3b}{2b-3a}$

Answers *a.* *b.*

c. *d.*

Section 5.7 Operations on Algebraic Fractions

Exercise 1 a. $\frac{2}{5} \cdot \frac{2}{3} =$ ***b.*** $\frac{2}{w} \cdot \frac{z}{3} =$

Exercise 2 a. $\frac{5}{4} \cdot \frac{8}{9} =$ ***b.*** $\frac{5}{2a} \cdot \frac{4a}{9} =$

Exercise 3 a. $\frac{7}{3} \cdot 12 =$ ***b.*** $\frac{7}{x} \cdot 4x =$

Exercise 4 a. $\frac{8}{3} \div \frac{2}{9} =$ ***b.*** $\frac{4a}{3b} \div \frac{2a}{3} =$

Exercise 5 a. $\frac{2}{3} \div 4 =$ ***b.*** $\frac{2}{3y} \div 4y =$

Exercise 6 a. $\frac{13}{6} - \frac{5}{6} =$ ***b.*** $\frac{13k}{n} - \frac{5k}{n} =$

Exercise 7 Complete each multiplication.

a. $\dfrac{-5}{b}a = \dfrac{-5}{b} \cdot \dfrac{a}{1} =$

b. $6x\left(\dfrac{2}{x^2 - x}\right) = \dfrac{6x}{1} \cdot \dfrac{2}{x^2 - x} =$

Exercise 8 Divide $\dfrac{6ab^2}{2a+3b} \div 4a^2b.$

Answer:

Exercise 9 Add $\dfrac{2n}{n-3} + \dfrac{n+2}{n-3}.$

Answer:

Exercise 10 Evaluate the sum in Example 10 for $x = -9$. How do you know that the answer will be -2 without carrying out any calculations?

Answer:

Exercise 11 Subtract $\dfrac{3}{x^2+2x+1} - \dfrac{2-x}{x^2+2x+1}.$

Answer:

Section 5.8 Least Common Denominators

Exercise 1 Bruce drove for 30 miles in the city at an average speed of 40 miles per hour. Then he drove 105 miles on the highway at an average speed of 70 miles an hour. Use the formula above to find his average speed for the entire trip.

Answer:

Exercise 2 Add $\frac{5x}{6} + \frac{3-2x}{4}$

Answer:

Exercise 3 Follow the steps below to subtract $\frac{2}{x+2} - \frac{3}{x-2}$.

Answer:

Step 1 Find the LCD:

Step 2 Build each fraction.

Step 3 Subtract like fractions.

Step 4 Reduce if possible.

Exercise 4 Find the LCD for $\frac{3}{4x^2} + \frac{5}{6xy}$.

Answer:

Exercise 5 Simplify $\dfrac{\frac{2x}{y^3}}{\frac{x}{3y}}$.

Answer:

Exercise 6 Evaluate the complex fraction in Example 4 for $x = 8$. Explain how you know that the answer is $\frac{1}{4}$ without doing a lot of calculation.

Answer:

Exercise 7 Follow the steps below to simplify $\dfrac{1 + \frac{b}{a}}{1 - \frac{bc}{ad}}$.

Answer:

Step 1 Find the LCD of $\frac{b}{a}$ and $\frac{bc}{ad}$.

Step 2 Multiply *each term* of the numerator and denominator of the complex fraction by the LCD from Step 1.

Step 3 Reduce your result if necessary.

Section 5.9 Equations with Algebraic Fractions

Exercise 1 A new health club opened up, and the manager kept track of the number of active members over its first few months of operation. The equation below gives the number, N, of active members, in hundreds, t months after the club opened.

$$N = \frac{10t}{4+t^2}.$$

The graph of this equation is shown in Figure 5.11.

a. Use the equation to find out in which months the club had 200 active members.

b. Verify your answers on the graph.

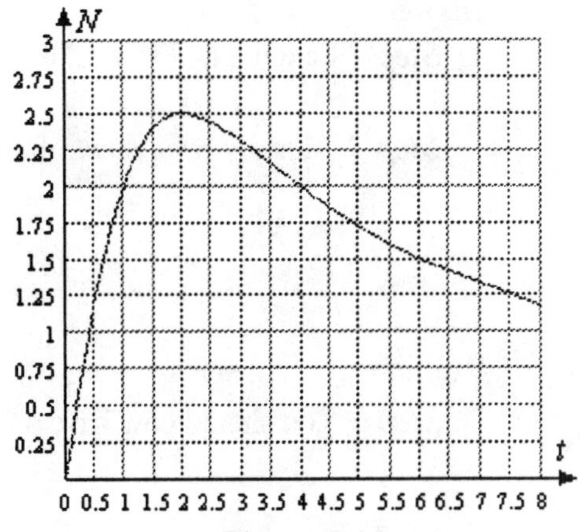

Figure 5.11

Answer:

Exercise 2 Solve $\frac{x^2}{2} + \frac{5x}{4} = 3$ by first clearing the fractions.

Answer

Exercise 3 Solve $\dfrac{1}{x-2}+\dfrac{2}{x}=1.$

Answer:

Step 1 Find the LCD for all the fractions in the equation: LCD = ________________

Step 2 Multiply *each term* of the equation by the LCD, and simplify.

$$(\qquad\qquad)\frac{1}{x-2}+(\qquad\qquad)\frac{2}{x}=1(\qquad\qquad)$$

Step 3 You should now have an equation with no fractions. Solve as usual.

Exercise 4 Follow the steps below to solve $\dfrac{15}{x^2-3x}+\dfrac{4}{x}=\dfrac{5}{x-3}.$

Answer:

Step 1 Factor each denominator and find the LCD for the fractions in the equation.

LCD =

Step2 Multiply *each term* of the equation by the LCD, and simplify.

$$(\qquad\qquad)\frac{15}{x^2-3x}+(\qquad\qquad)\frac{4}{x}=\frac{5}{x-3}(\qquad\qquad)$$

Step 3 You should now have an equation with no fractions. Solve as usual: Begin by using the distributive law to remove parentheses.

Step 4 Check for extraneous solutions.

Exercise 5 Solve $\dfrac{w}{q} = 1 - \dfrac{T_c}{T_h}$ for q.

Answer:

Exercise 6 The city reservoir was completely emptied for repairs and is being refilled. Water flows in through the intake pipe at a steady rate that can fill the reservoir in 120 days. However, water is also being drained from the reservoir as people use it, so the reservoir actually takes 150 days to fill. After the reservoir is filled, the intake pipe is turned off, but people continue to use water at the same rate. How long will it take to drain the reservoir dry?

Answer To solve the problem, compute the flow rate for each pipe, intake and outflow, separately.

a. Let d stand for the number of days for the outflow pipe to drain the full reservoir. Write an expression for the fraction of the reservoir that is drained in one day: __________ This is the rate at which water leaves the reservoir.

b. Now imagine that the intake pipe is open, but the outflow pipe is closed. Write an expression for the fraction of the reservoir that is filled in one day: __________ This is the rate at which water enters the reservoir.

c. Consider the time period described in the problem, when both pipes are open. Fill in the table.

	Flow rate	*Time*	*Fraction of reservoir*
Water entering			
Water leaving			

d. Write an equation:

$$\left(\begin{matrix}\text{Fraction of reservoir}\\ \text{filled}\end{matrix}\right) - \left(\begin{matrix}\text{Fraction of reservoir}\\ \text{drained}\end{matrix}\right) = \text{One whole reservoir}$$

e. Solve your equation.

Chapter 6 More About Exponents and Radicals

Section 6.1 Laws of Exponents

Exercise 1 Use the laws of exponents to simplify.

a. $(4x^4)(5x^5)$ *b.* $\dfrac{8x^8}{4x^4}$ *c.* $\dfrac{3^2a^3(3a^2)}{2^3a(2^3a^4)}$

Answers *a.* *b.* *c.*

Exercise 2 Write $(y^2)^5$ as a repeated product and apply the first law of exponents to simplify it.

Answer: $(y^2)^5 =$

Exercise 3 *a.* $(5^3)^6 =$

b. $(y^4)^2 =$

Exercise 4 Simplify each expression using the laws of exponents. Then use your calculator to verify that your answers are correct.

a. $(5^4)^2 =$

b. $(5^4)(5^2) =$

Exercise 5 Simplify $(6q^6)^2$.

Exercise 6 Recall that $-x$ can be interpreted as $-1 \cdot x$. Use this idea to simplify each expression if possible.

a. $-x^4$

b. $(-x)^4$

c. $-x^5$

d. $(-x)^5$

Exercise 7 Simplify $\left(\dfrac{n^3}{k^4}\right)^8$.

Exercise 8 Follow the suggested steps to simplify $\left(\dfrac{-2ab^4}{3c^5}\right)^3$.

Step 1 Apply the fifth law: raise numerator and denominator to the third power.

Step 2 Apply the fourth law: raise each factor to the third power.

Step 3 Apply the third law: simplify each power of a power.

Exercise 9 Follow the suggested steps to simplify $3x(xy^3)^2 - xy^3 + 4x^3(y^2)^3$.

Step 1 Identify the three terms of the expression by underlining each separately.

Step 2 Simplify the first term: apply the fourth law, then the third law.

Step 3 Simplify the last term: apply the third law.

Step 4 Add or subtract any like terms.

Section 6.2 Negative Exponents

$2^4 = 16$		$5^4 = 625$	
	Divide by 2		Divide by 5
$2^3 = 8$		$5^3 = 125$	
	Divide by 2		Divide by 5
$2^2 = 4$		$5^2 = 25$	
$2^1 =$		$5^1 =$	
$2^0 =$		$5^0 =$	
$2^{-1} =$		$5^{-1} =$	
$2^{-2} =$		$5^{-2} =$	
$2^{-3} =$		$5^{-3} =$	

Exercise 1 Answer the following questions about your lists:

a. What did you find for the values of 2^0 and 5^0?

b. If you make a list with another base (say, 3, for example), what will you find for the value of 3^0?

c. Can you explain why this is true?

d. Compare the values of 2^3 and 2^{-3}. Do you see a relationship between them?

e. What about the values of 5^2 and 5^{-2}? Try to state a general rule about powers with negative exponents.

f. Use your rule to guess the value of 3^{-4}.

Exercise 2 Write each expression without exponents.

a. -6^2 *b.* 6^{-2} c. $(-6)^{-2}$

Answers: *a.* *b.* *c.*

Exercise 3 Write each expression without negative exponents.

a. $4t^{-2}$ *b.* $(4t)^{-2}$ *c.* $\left(\frac{x}{3}\right)^{-4}$

Answers *a.* *b.* *c.*

Exercise 4 Simplify by using the first or second law of exponents.

a. $3^{-3} \cdot 3^{-6}$ *b.* $\dfrac{b^{-7}}{b^{-3}}$

Answers *a.* *b.*

Exercise 5 Write without negative exponents and simplify.

a. $\dfrac{1}{15^{-2}}$ *b.* $\dfrac{3k^2}{m^{-4}}$

Answers *a.* *b.*

Exercise 6 Simplify by using the third or fourth law of exponents.

a. $(3y)^{-2}$ *b.* $(a^{-3})^{-2}$

Answers *a.* *b.*

Exercise 7 Explain the simplification of $\dfrac{(3z^{-4})^{-2}}{2z^{-3}}$ shown below. State the law of exponents or other property used in each step.

Explanation:

Solution

$$\frac{(3z^{-4})^{-2}}{2z^{-3}} = \frac{3^{-2}(z^{-4})^{-2}}{2z^{-3}} \qquad (1)$$

$$= \frac{3^{-2}z^{8}}{2z^{-3}} \qquad (2)$$

$$= \frac{3^{-2}z^{8-(-3)}}{2} \qquad (3)$$

$$= \frac{z^{11}}{3^{2}\cdot 2} = \frac{z^{11}}{18} \qquad (4)$$

Section 6.3 Scientific Notation

Exercise 1 Compute each product.

a. 1.47×10^5 *b.* 5.2×10^{-2}

Answers *a.* *b.*

Exercise 2 Fill in the correct power of 10 for each factored form.

a. $0.00427 = 4.27 \times$ ________ *b.* $4800 = 4.8 \times$ ________

Exercise 3 Write each number in scientific notation.

a. The largest living animal is the blue whale, with an average weight of 120,000,000 grams.

b. The smallest animal is the fairy fly beetle, which weighs about 0.000005 grams.

Answers *a.* *b.*

Exercise 4 Perform the following calculations on your calculator. Write the results in scientific notation, and round to two decimal places.

a. $6{,}565{,}656 \times 34{,}567$ *b.* $0.000000123 \div 98{,}765$

Answers *a.* *b.*

Exercise 5 Use scientific notation to find the quotient

$$0.00000084 \div 0.0004$$

Answer:

Step 1 Write each number in scientific notation:

Step 2 Combine the decimal numbers and the powers of 10 separately:

Section 6.4 Properties of Radicals

1. Is it true that $\sqrt{a+b} = \sqrt{a} + \sqrt{b}$?

Examples

(i) Does $\sqrt{9+16} = \sqrt{9} + \sqrt{16}$?

(ii) Does $\sqrt{2+7} = \sqrt{2} + \sqrt{7}$?

2. Is it true that $\sqrt{a-b} = \sqrt{a} - \sqrt{b}$?

Examples

(i) Does $\sqrt{100-36} = \sqrt{100} - \sqrt{36}$?

(ii) Does $\sqrt{12-9} = \sqrt{12} - \sqrt{9}$?

3. Is it true that $\sqrt{ab} = \sqrt{a}\sqrt{b}$?

Examples

(i) Does $\sqrt{4 \cdot 9} = \sqrt{4} \cdot \sqrt{9}$?

(ii) Does $\sqrt{3 \cdot 5} = \sqrt{3} \cdot \sqrt{5}$?

4. Is it true that $\sqrt{\frac{a}{b}} = \frac{\sqrt{a}}{\sqrt{b}}$?

Examples

(i) Does $\sqrt{\frac{100}{4}} = \frac{\sqrt{100}}{\sqrt{4}}$?

Exercise 1 Decide whether the statement is true or false, and then verify with your calculator.

a. $\sqrt{12} = \sqrt{4}\sqrt{3}$ *b.* $\sqrt{12} = \sqrt{8} + \sqrt{4}$

Answers *a.* *b.*

Exercise 2 Simplify $\sqrt{75}$.

Answer:

Exercise 3 Find the square root of each power.

a. $\sqrt{y^4}$ *b.* $\sqrt{a^{16}}$

Answers: *a.* *b.*

Exercise 4 Simplify $\sqrt{72u^6v^9}$.

Answer:

Exercise 5 Write each expression as a single term.

a. $13\sqrt{6} - 8\sqrt{6}$ *b.* $5\sqrt{3x} + \sqrt{3x}$

Answers *a.* *b.*

Section 6.5 Operations with Radicals

Exercise 1 Find the surface area of the Sun Pyramid in Mexico. The pyramid is 210 feet high and has a square base 689 feet on each side.

Answer:

Exercise 2 Find the product and simplify: $\left(3x\sqrt{6x}\right)\left(y\sqrt{15xy}\right)$

Answer:

Exercise 3 Multiply $2a\left(2\sqrt{3}-\sqrt{a}\right)$

Answer:

Exercise 4 Multiply $3\sqrt{x}\left(\sqrt{2x}-3x\right)$

Answer:

Exercise 5 Expand and simplify $\left(\sqrt{2}-2\sqrt{b}\right)^2$.

Answer:

Exercise 6 Add $\dfrac{\sqrt{x}}{3} + \dfrac{\sqrt{2}}{x}$

Answer:

Exercise 7 Solve $(2x+1)^2 = 8$

Answer:

Exercise 8 Simplify $\dfrac{3ab\sqrt{75a^3b}}{\sqrt{6ab^5}}$.

Answer:

Exercise 9 Rationalize the denominator of $\dfrac{8}{\sqrt{x}}$.

Answer:

Section 6.6 Equations with Radicals

Exercise 1 Evaluate the formula to find the speed the cars must reach on each roller coaster in order to stay on the track. Because the vertical loops used in roller coasters are not perfect circles, the total height of the loop is about 2.5 times its radius. (See Figure 6.7.)

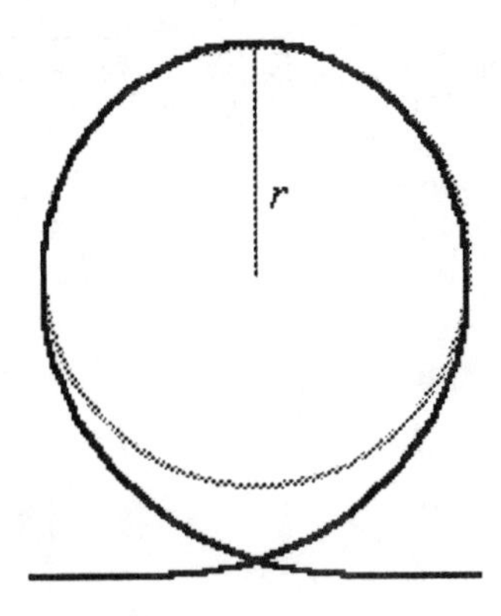

Figure 6.7

Answers: Viper:

Gash:

Shockwave:

Exercise 2 You would like your roller coaster to have a vertical loop that is 200 feet tall. How fast must the cars travel?

Answer:

Exercise 3 Calculate the maximum loop height possible for the roller coaster Fujiyama.

Answer:

Exercise 4 Solve $\sqrt{x-6} = 2.$

Answer:

Exercise 5 Solve $\sqrt{x-3} + 5 = x.$

Answer:

Exercise 6 Solve $3\sqrt[3]{4x-1} = -15.$

Answer:

Homework 6.6

13. a. Complete the table of values and graph $y = \sqrt{x-4}$ on the grid.

x	4	5	6	10	16	19	24
y							

b. Solve $\sqrt{x-4} = 3$ graphically and algebraically. Do your answers agree?

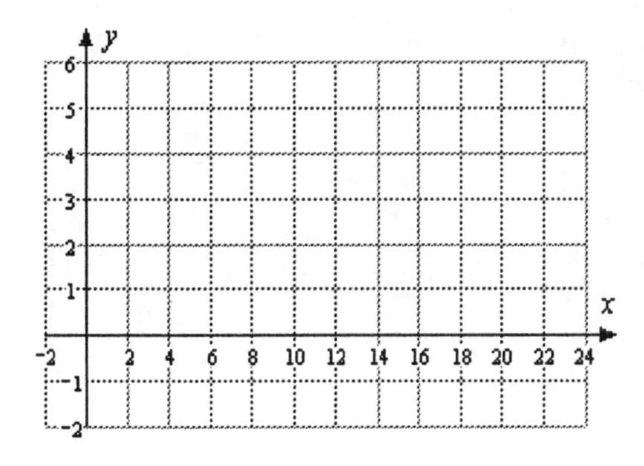

14. a. Complete the table of values and graph $y = 2 - \sqrt{x}$ on the grid.

x	0	1	4	8	12	18	24
y							

b. Solve $2 - \sqrt{x} = -1$ graphically and algebraically. Do your answers agree?

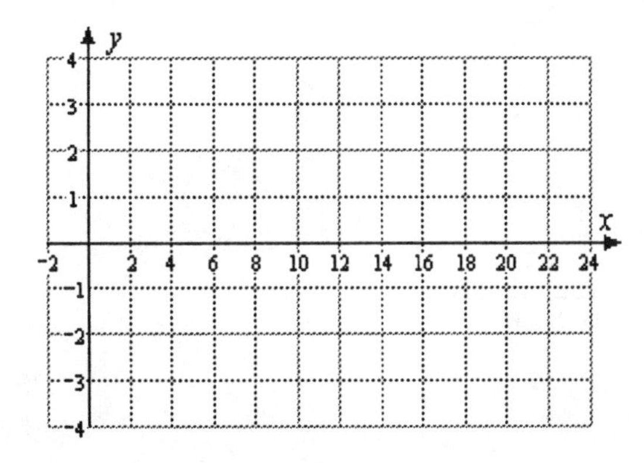

15. a. Complete the table of values and graph $y = 4 - \sqrt{x+3}$ on the grid.

x	-3	-2	0	1	4	8	16
y							

b. Solve $4 - \sqrt{x+3} = 1$ graphically and algebraically. Do your answers agree?

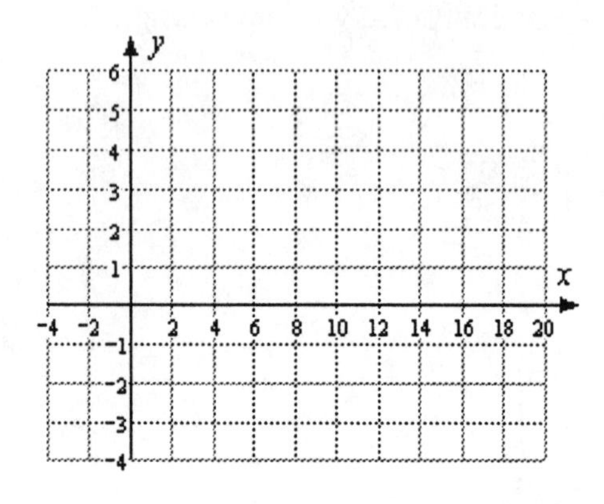

16. a. Complete the table of values and graph $y = 2 + \sqrt{3x-6}$ on the grid.

x	3	4	5	8	10	16	20
y							

b. Solve $2 + \sqrt{3x-6} = 8$ graphically and algebraically. Do your answers agree?

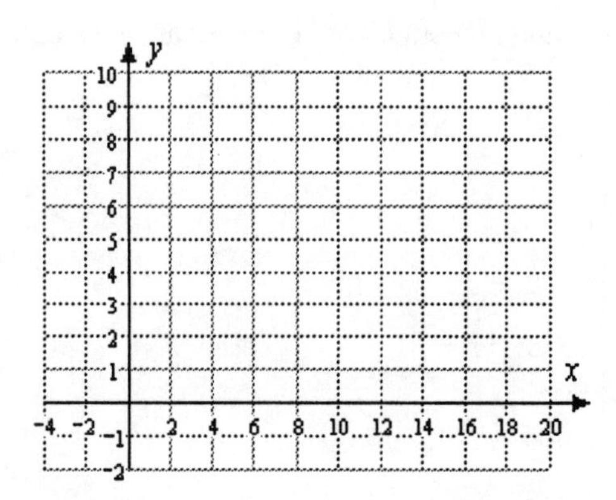